The Principles of Housing

T0186328

The Principles of Housing is an engaging and discursive introduction to the key topics within housing studies. Whereas many books get bogged down in country-specific policy or small innovations, this book argues that the fundamental concepts of what we call housing are relatively stable and unchangeable. By focusing on universal principles, the book provides an introduction to housing that can be used by students world-wide.

The book consists of a series of short chapters relating to the key issues of housing, such as borrowing, choice, finance, government, need, reform and welfare. Each chapter is designed to be a starting point for a wider conversation, with discussion questions and a number of think pieces and international case studies to help students connect these general principles to their own surroundings.

Written by renowned housing expert Peter King, *The Principles of Housing* succeeds in being accessible and engaging without shying away from the complexities of housing issues. The book will be invaluable to students on housing-related courses across finance, real estate, planning, development, politics and sociology subjects. The book will also be useful to housing professionals and policymakers aiming to expand their understanding of housing issues.

Peter King has 25 years' experience of teaching housing issues and is the author of 18 books. He is currently a Reader in Social Thought at De Montfort University, UK.

Peter King's new book, drawing on his extensive experience of housing teaching and scholarship, offers a very distinctive and valuable contribution. It sets out to explore and explain core principles of housing and how we may actually understand housing in a conceptual sense. The book's accessible style, and the very wide range of topics and concepts covered across its many short chapters, make it an ideal introduction and teaching resource directly relevant to a wide range of courses at all levels. The book's engaging and thought provoking arguments will equally appeal to housing researchers and practitioners. This text undoubtedly achieves its aim of acting as a catalyst for further conversations about what housing is and what and who it is for.

John Flint, Professor of Town and Regional Planning,
University of Sheffield, UK

Peter King provides a fascinating insight into the principles underlying discussions of housing. Written in a highly engaging, yet extremely thoughtful style, the book offers an original and accessible analysis of key concepts used in housing debate. The book will have a dual appeal, in offering both a broad introduction to, and concise, but more detailed analysis of fundamental issues. The book is thus required reading for anyone with an interest in understanding this often complex and always contested field. It fully deserves to become a core text in contemporary housing studies.

Tony Manzi, University of Westminste, UK

The Principles of Housing

Peter King

Routledge
Taylor & Francis Group

LONDON AND NEW YORK

First published 2016
by Routledge
2 Park Square, Milton Park, Abingdon, Oxon OX14 4RN

and by Routledge
711 Third Avenue, New York, NY 10017

Routledge is an imprint of the Taylor & Francis Group, an informa business

British Library Cataloguing in Publication Data
A catalogue record for this book is available from the British Library

Library of Congress Cataloging in Publication Data
King, Peter, 1960-
The principles of housing / Peter King.
pages cm
Includes bibliographical references and index.
1. Housing. 2. Public housing. 3. Housing policy. I. Title.
HD7287.K5588 2016
363.5--dc23
2015021933

ISBN: 978-1-138-93941-7 (hbk)
ISBN: 978-1-138-93942-4 (pbk)
ISBN: 978-1-315-67498-8 (ebk)

Typeset in Bembo
by Integra Software Services Pvt. Ltd.
Printed in Great Britain by Ashford Colour Press Ltd, Gosport, Hants

Contents

PART 7
Buildings **167**

Preface

This book might be described as my 'greatest hits'. It consists of most of the ideas, concepts and discussions that I have used in my teaching over the last 25 years. And these conversations are the best way I have found to describe and discuss housing issues so far, and it may be that I won't ever be able to better it. This might well provide good reason to pity me, and I am certainly not claiming that this is the best that can be said. Others have certainly said these things in a different manner and there will doubtless be others in the future. So I feel very fortunate to have this opportunity of putting my particular views forward.

This book began with a very speculative and perhaps even frivolous question: what sort of housing book would I most like to write if there were no restrictions placed on me? After some thought I determined that the book would consist of many short chapters that covered the core principles of housing, written in such a manner that it could apply to Britain, the US, Europe and pretty much anywhere else and could have applied to housing in 1990 and might equally be relevant in 2040. I played with this idea and enjoyed doing so, but never thought that such an eccentric project would lead anywhere. But I was soon able to put together what seemed to be a viable book that developed many of the ideas in my earlier work, *Understanding Housing Finance*. And I was surprised and delighted when the book was received so enthusiastically. So I have three sets of people to thank who have helped me while I have worked on this book.

First, I must thank Helena Hurd, Sade Lee and their colleagues at Routledge for their enthusiasm and incredibly positive support for the project. Helena's advice has been crucial to improving the book. I am also grateful to the anonymous reviewers for their comments and constructive criticism.

This brings me on to the group that perhaps I owe most to for this project. These are the students who have taken my classes over the last 25 years, particularly my modules, *Housing Markets and the State* and *Housing, Health and Social Policy*. This is where most of the ideas in this book have been trialled, honed and argued over. I feel honoured to have worked with such honest and open human beings who take housing issues so seriously, but without any dogma or preconceptions. I have certainly learnt more from them than they have from me.

Finally, I turn to those who mean the most to me: my wife, B, who has to put up with endless hours of me worrying over projects, going from despair to elation and seemingly never learning. She meets my daftness with patience and good humour and never fails to encourage and support. My daughters, Helen and Rachel, now easily outstrip me intellectually and this is a joy as a parent, but chastening as a hoary old academic who has long since realised that the world wasn't going to change because of anything I did or said. But they still listen and humour me (and proofread my books). All in all, I am truly the most fortunate man in the world.

Peter King
June 2015

Introduction

Most books on housing, whether they are introductory or more advanced, are really concerned with aspects of housing policy in one country or perhaps with comparing policy in a few countries. What these books do not tend to do is to deal with the core principles of housing. Indeed, there seems to be very little recognition that there are such things as core principles with regard to housing. There is very little discussion of what housing means in a conceptual sense. There are many books on what it means to educate a child or a young adult, but not many on what it means to house a household. This may be because the subject matter is rather different, but it is still noteworthy that much of the work on housing is focused on policy, with little that is more general or conceptual. This book is an attempt to remedy that by offering discussions of core issues in a manner that does not relate solely to one particular system at one particular time.

However, the book is aimed as an introductory text rather than as a high-falutin' monograph. What I seek to do here is consider the main principles of housing in a manner that is accessible and open to those who are studying housing at a range of levels, including those just starting out.

While I have chosen to call this book *The Principles of Housing*, it does not claim to say everything about housing. It can, I hope, be used as a text to support a range of courses on housing. But I also see it as providing a jumping-off point for discussion, even if the purpose of the discussion is to completely disagree with the contents of a chapter. This is not written as the last word, even though it is very probably the best manner in which I am able to say these things. What I provide is a series of critical descriptions of what housing is and how and why it is produced, sustained and used. I have tried not to take anything for granted, such that someone with little knowledge of housing issues can understand the book and make use of it. But I have also tried to make it sufficiently distinctive and interesting to catch experts in the field as well. The book is then an introduction to housing, in that it considers the key issues, but it is by no means a conventional introduction in that its focus is general and conceptual.

While the work can be seen as introductory, I also have a larger purpose. What I aim to show here is that, despite the rhetoric of constant change and apparent permanent flux of new initiatives, all with their own new jargon, the reality is that most of what we actually take to include housing has not changed much and will not change much either in the future. It is therefore entirely possible to say what housing is.

Much of what we call housing is relatively stable and does not change. We tend to lose sight of this stability by focusing instead on small innovations and developments which enrapture us for a time before we turn to the next fad. With each fad we maintain our enthusiasm, hoping not to be caught out by the latest thing, not be seen as cynical, divisive, naïve or, God forbid, old-fashioned. The fact that these little innovations follow each other so quickly presents the illusion of constant change, and that we have to be constantly running to keep up. However, the real basis of housing remains unchanged, and we need to appreciate this. This is the main message and purpose of this book: to state what is here and what remains.

Much of my housing teaching in recent years has been based on my book, *Understanding Housing Finance*, which has been published in two editions in 2001 and 2009. The first book was rather conventional, providing a historical overview and a tenure-by-tenure guide. To my frustration, I found that the book quickly became out of date. The second edition was an attempt to write a book that would not date so readily and so focused more on concepts. This meant that there was a lot less detail on policies, but I was rather more satisfied with the book and it has been more successful as a teaching aid.

This book began as a third edition, but it has turned into something rather different. This change has come about because much of my teaching is no longer really on housing finance, but is now much broader. This book, therefore, seeks to mirror this more general shift. So, instead of a third edition, there is this rather different book.

But this shift in focus has also been driven by changes in my style of teaching. Increasingly I have shifted from presenting material formally and instead have adopted a teaching style based on discussion and debate. This involves a degree of extemporisation such that no two classes are the same. Much depends on how the students respond to my questions and comments and what they are prepared to contribute (and yes, there have been some disasters).

Much of my teaching has been with small groups – seldom more than 15 – consisting largely of mature students who are seeking to gain either a professional or academic housing qualification or both. Most of these students work full time for housing organisations and so have considerable knowledge of day-to-day housing practices and how to deal with people in severe need. What they often lack is the bigger picture that allows them to see how policies at the local level connect up with the national situation and how various parts of the system link together. Increasingly I came to see my job as to provide this overall view and I chose to do it through discussion and debate rather than by a more didactic approach. My aim then was to develop a conversation.

But using this approach means that I cannot hide behind anything and I tend not to go into a class with overmuch preparation. I will perhaps have a lesson plan consisting of 100 or so words and the occasional very brief handout that contains important information that I want to make sure the students have grasped. Other than that, I rely on my wits and a couple of marker pens and see where we end up. No two classes are the same, and I am never sure what might happen. It is tremendous fun. Of course, I would not have been able to do this as a new teacher and I would not particularly recommend it to others. But, as the author of 16 books and a teacher of 25 years' experience, I feel that I can now do it and I find that it suits both the students and myself.

This approach to teaching means that there are no barriers between students and teacher. There is no technology, no literature, just them and me. There is a directness based on a learned ability on my part to distil what I have gleaned in my research, reading and past teaching experience. This text is an attempt to replicate this sense of an unmediated experience in book form. It is obviously only an approximation and, of course, you only get my side of the discussion, but I hope that the informal conversational nature will give a flavour of it.

But this means that I have left behind many of the usual academic conventions in favour of a directness and simplicity of presentation. In particular, the text is not cluttered with references and it is very light on facts and figures. Each short chapter seeks to make a statement that both informs and acts as a platform for further study and discussion.

As I have stated, the aim of this book is to be rather more general than other books on housing. Accordingly, there is very little in the book about specific policies or what happens in one particular country. Instead, my concern has been with what I see as the principles of housing. It is about those things that do not really change and which tend to be shared across housing systems. My hope is that the book will therefore have a reasonably long shelf life and also have its uses in a large number of countries, rather than just focusing on one system at one point in time.

In some ways this can be seen as a rather old-fashioned project. I am claiming to say something definitive – what are the principles of housing – and to do it in a manner that is conversational rather than strictly adhering to academic conventions. I hope that I am not being deliberately wilful in presenting the material in this manner. My aim has been to provide something that is readable while also being definitive. This is not because I think this is the only way of describing housing phenomena, but it is the best way that I can. It is *my* definitive position, and it is of course for others to determine if that is in any way valuable. Other writers would deal with these issues differently, whether in terms of content, analysis or style, and I have no desire to suggest that others' approaches are not valid or important. However, after 25 years, this is what I can come up with.

This book, though, is not meant to be backward looking. It is not intended to be a testament or a monument to anything. It is not a mere record of what I used to get up to. I hope that it makes a serious contribution to housing studies and teaching. The book focuses on principles rather than facts and makes the very serious point that much of housing does not change and that the fundamentals remain pretty much the same. What I have sought to do is to reduce housing to its core elements. This is not because I consider housing to be a simple matter, nor do I want to suggest that there is a pattern or model that can be applied generally. The principles I discuss here are many, some are more important than others, and putting them all together creates a complexity that cannot be readily summed up. That is why there is no attempt in the book to pull it all into a coherent narrative. So I do not think that it is possible to sum up housing by a few statements. If asked what the principles of housing are, I would give a long list based on the contents page of this book. So what I am saying is that housing can be reduced to a relatively large number of complex statements. Just because something does not change does not make it simple or easy to understand.

It is perhaps the height of arrogance to try to tell readers how to use a book: if you have bought it you can do what you like with it! However, perhaps a few words here might be useful. One way to view the book is to imagine that the word 'Discuss' appears at the end of every chapter. The aim is to stimulate discussion as well as to provide accurate information. So the chapters are seen as merely starting points, opening up the issues and providing the opportunity for further development. I have provided a number of discussion points at the end of each chapter to start things off. There are also a number of think pieces scattered throughout the book, which are linked to particular chapters. These pieces are intended, too, to spark off discussion.

The book, of course, can be read from cover to cover, and I hope that it would prove enjoyable and useful if one did so. However, it is rather intended as the sort of book to dip into for something specific. The chapters have been grouped together into a number of themes, but they can of course be read as separate stand-alone pieces, and that is how they have been written. The index provides more specific help in searching for issues. Each chapter includes links to related chapters and some suggestions for further reading. The texts mentioned are often not the most recent, but I consider them to be the most useful and interesting. However, an invaluable general reference for the discussions in this book is the *International Encyclopedia of Housing and Home*, edited by Susan Smith and colleagues. This contains discussions on all the topics raised here, often linked to specific regions or countries. This would be a good starting point for continuing study on the principles of housing outlined here.

Further reading

Smith, S. J., Elsinga, M., Fox O'Mahony, L., Eng, S. O. and Wachter, S. (2012) (Eds): *International Encyclopedia of Housing and Home*, Oxford, Elsevier.

Part 1
The basics

1 Housing and home

What do we mean when we say 'housing'? This sounds like a very silly question, because we all think we know. After all, we live in the stuff. But is it really as straightforward as we might think?

The first issue is that the word 'housing' can be used as both a noun and a verb. It can be used to describe a thing but also an activity. So we can suggest that housing is both a collection of dwellings *and* the activity of providing, managing and maintaining a collection of dwellings. Housing, then, is both a thing and something that people do.

But note also, then, that, in defining housing, we used the word 'dwelling'. We did this to make it clear that we are not merely referring to 'houses' but to a range of different forms of accommodation. Most definitions of dwelling refer to it as a building or structure. Again, we cannot presume it is a fixed building, but it might also include a mobile home or caravan, a tent or the use of a natural structure such as a cave.

An alternative way of defining housing might be to refer to it as shelter, and this has the advantage of having no specific connotation to a particular built form or structure. But shelter is also used in different ways depending on the context. The word 'shelter' can be suggestive of an existential need. Shelter is something that helps to keep us alive by providing the basics of warmth, security and protection from the elements. But shelter can also refer to a temporary or immediate respite from the elements. We take shelter from the rain under a tree or in a shop doorway. This form of shelter is not intended to provide us with a permanent solution.

But housing is more than just building. We are also concerned with who can live in it and whether it is sustainable as a permanent residence. We are interested in how it is paid for and whether all households can afford access to housing that is of decent quality. Hence, housing is also an activity that concerns itself with these things. And, because these activities involve skills, judgement and analysis on the part of those tasked with them, we can also talk of housing as being a profession, or perhaps better as a series of professions that includes developers, housing management and estate and residential letting agents, as well as those involved in planning and maintenance.

So, for many of the chapters in this book, we shall be referring to housing as a verb, as an activity that is concerned not just with things, but with the activities that create and sustain those things. Hence much of what interests us is to do with finance, legislation, regulation and the institutions that undertake these tasks.

But there is still more to housing than this. We need the structures to shelter us and we need them to be well built, well maintained and affordable. But most of all we have to be able to use the structures. Housing can all too easily be seen as an end in itself. We often hear people state that we need more housing. But what do we need it for? The answer to this is that housing is what we use in order to make homes for ourselves. A home is a place of nurture and a store of memories. It is a place where we can be intimate with those we love and who love us. It is where we raise children and also a place where we can be ourselves, secure in a private space that we control and that we are able to use as we see fit. Housing therefore allows us to live as private beings, to nurture the next generation and to live in security, peace and comfort.

So much of what matters about housing can only take place once the door is firmly shut and the rest of the world is excluded. It is this exclusion that allows us to make and maintain a home, and this is what housing is really about. All the technicalities, the policies and the financial models exist only to allow households to make a home.

But there has been something of an unfortunate shift in recent years, in that we do not talk so much about housing or dwellings any more, but rather about homes. The housing and building professions, as well as politicians, now commonly use 'home' instead of 'house' when they refer to physical structures. Social landlords manage and build homes and not dwellings or houses. The reason for this is clearly that home is a warmer, more evocative concept, which converts a brick box into something with a much stronger emotional resonance. Accordingly, when we discuss those lacking a dwelling, we call them 'homeless' to emphasise the full import of what they are suffering and the full possibility of its redemption. 'House' is a cold and empty word, which becomes inhabited and warm when translated into 'home'.

Speaking only of homes adds a greater significance to what housing and building professionals are doing: they are not building or managing brick boxes, but creating something warm and welcoming to residents. But, in doing so, are we taking something away from the concept of home itself? It is no longer a place of nurturing and comfort, but now a physical structure.

Our home is, we hope, a store of memories. It is a place that is comfortable and comforting and provides us with security. It is a both a refuge and a nest, and each one is unique to those who live there. Thus the misuse of the word 'home', so that it refers to physical structures, diminishes what it ought to signify. The aim of adding significance to the quotidian tasks of managing and building reduces the idea of 'home' to something empty and cold.

This misuse is significant in another sense, namely that it implies that homes are 'made' by those other than the household. Homes, we are now led to believe, are made by professionals, ready-made for people to live in. The household no longer has to create or make the home; the work has been done for them. This situation has several consequences. First, because home is ostensibly created by professionals, this implies that no effort is needed on the part of the household. The suggestion is that homemaking is easy to achieve and can be readily done for us. Second, this view carries the apparent belief that home is transient. Building homes implies that we move from home to home and do not take it with us. Third, this idea implies the standardisation of homes according to professionals' understanding of their clients' needs and aspirations. The result is the provision of identikit homes, based on standard design briefs and models. This creates an increasing homogeneity of styles aimed to fulfil standardised purposes. We need only think of terms such as 'starter home' and 'executive home' to see this process of standardisation. Fourth, this will tend to impersonalise the notion of home and dwelling more generally: it becomes a commodity that is bought and sold and treated as such. Housing is commodified according to economic rather than human values. Lastly, but implicit to all the above, this notion of home assumes the professionalisation of the role of homemaking: homes can only be made by others, by 'the experts'. Professionals tell us what we need or, in other words, they actually deem to tell us what home is.

So we should be careful with words and we should certainly insist on a separation between housing and home: they are two different things. This, though, does not in any way diminish the importance of housing. Separating it from the idea of home does not mean that there is nothing left for professionals to do. Instead, we need to remember that the use of housing – and therefore the need to create it, manage it, finance it and so on – is ubiquitous. We all need it and will continue to need it. Our housing may differ in terms of style and size, from an igloo to a twenty-first-floor flat, but it is still housing, and the process that creates it is also housing.

Housing fulfils a basic need or needs, for shelter, security, autonomy and so on. We most certainly cannot manage without it. But we are not merely satisfied by the basics. We have particular standards, expectations and aspirations. We expect heating, running water, space, and if these are not present then the house is determined to be unfit or uninhabitable. We want something much more complex than basic shelter. We actually want a sophisticated machine, or rather a collection of machines, which all work for us seamlessly and without causing us any trouble. But this takes money, a legal system, regulations and planning, and only when these are present can we truly forget the complexity of the activity of housing and properly call it our home.

Discussion points

1   What is home?
2   How does housing differ from home?
3   Is the distinction between housing as a noun and a verb an important one?

See also

Chapter 13 Desire
Chapter 27 Boom and bust
Chapter 33 Development
Chapter 35 Architecture

Further reading

Bachelard, G. (1969): *The Poetics of Space*, Boston, MA, Beacon Books.
Blunt, A. and Dowling, R. (2006): *Home*, London, Routledge.
Clapham, D. (2005): *The Meaning of Housing: A Pathways Approach*, Bristol, Policy Press.
King, P. (2004): *Private Dwelling: Contemplating the Use of Housing*, London, Routledge.

2 Quality and access

Housing is one of the most important items that we human beings need. There are many things that we would find difficult, if not impossible, to do without good quality housing. We might find it hard to get and keep a job, to learn, to maintain our health, to vote, to claim benefits that we are entitled to and to initiate and maintain stable relationships.

But, just because something is important, this does not mean that it is always available. Like most commodities, housing comes with a price tag attached. If we want decent housing, we have to pay for it. It also follows, broadly speaking, that the better the standard of housing we want, the more it will cost us. Therefore, as standards rise, so does the cost.

One of the most important issues, then, is how we can afford the sort of housing that we want. We could say that this is simply a case of matching up our income with our aspirations and expectations and buying the best dwelling we can afford. This may be an option for those on reasonable incomes, but not for those on low incomes. Many households will lack sufficient income to provide them with a dwelling that meets their expectations. It may well be that they could find housing of some sort, but this might not be of a standard that they, or the society of which they are a part, find acceptable.

This implies that two issues are of supreme importance. The first is quality. We are not content with just any type of housing; we want good quality housing that allows us to live a civilised and healthy existence. We therefore require housing to a modern standard of amenity. This standard, of course, is a relative one, in that it depends on general expectations that exist here and now. It is no good saying that households elsewhere in the world manage with less or that our grandparents were brought up without central heating and modern appliances.

The second issue follows on from this and is about access. We might readily agree on what constitutes a good quality dwelling for us here and now. We can describe the particular amenities and standards that the modern dwelling should have. But that does not mean that everybody has such a dwelling. Many households might not be able to afford one.

There is a clear trade-off between quality and access, in that, generally speaking, the higher the quality, the fewer will be able to gain access to it.

Quality comes at a cost, and this limits access. There is, then, a gap that needs to be filled between the aspirations that people have for good quality housing and their ability to access it because of a lack of income. This gap is what the activity of housing is meant to fill.

Discussion points

1　Why does it matter what quality of housing we provide?
2　How do we know what constitutes good quality housing?
3　How do we ensure that everyone has access to good quality housing?

See also

Chapter 18 Fairness
Chapter 22 Sources of finance
Chapter 30 Government
Chapter 33 Development

Further reading

King, P. (2009): *Understanding Housing Finance: Meeting Needs and Making Choices*, 2nd edn, London, Routledge.

3　The past

It is difficult to discuss the history of housing in general terms because each country has developed their provision differently according to their own priorities, culture and political and social structures. However, it is possible to make some points that apply in many, if not exactly all, cases, and it is important to do this because it tells us something about the nature of housing. Despite the differences between countries, there are a number of factors that are common.

In some ways we can characterise the development of housing as periodic shifts between quantity and quality. In some periods there is a shortage of housing and so the emphasis is on increasing the quantity of housing available. At other times the emphasis is on housing quality, with concern about overcrowding, lack of amenities and poor standards. These standards tend to be relative rather than absolute, and so what might have been seen as eminently acceptable, or even desirable, in one period is considered a problem at a later time.

One of the common causes of shortage is migration, either internal within a country or external from other countries. In particular, many countries saw internal migration as a result of industrialisation leading to urbanisation. The development of new cities due to industrialisation acted as a magnet for millions of people seeking employment. The result of this unplanned shift in population was overcrowding and low quality development within cities, which in turn led to the spread of plague and disease due to unsanitary conditions and lack of infrastructure such as sewers and supplies of clean water. This urban growth occurred in the developed world in the late nineteenth and early twentieth centuries, but has occurred in the developing world consistently over the last 100 years as countries industrialise. As a consequence, the problems of unplanned housing development, with the allied problems of overcrowding and public health issues, are recurrent ones.

As an aside, we can make a general point here about the impact of technology on housing. As we have just mentioned, industrial production led to urbanisation and the growth of city living. But also other forms of technological change have had a similar impact, such as the development of railways and motorised transport, which allowed households to live further from work and to commute. In addition, while a brick box is a brick box, there have been considerable technological changes in a dwelling, such as heating systems and the use of machines such as fridges, microwaves, televisions and portable computing. All these technological innovations have had an impact on the nature of housing, having implications on space standards and the manner in which we use our dwelling.

The problems of overcrowding and unsanitary conditions led to calls for government intervention to improve the quality of housing and deal with the consequences for public health. As a result, much of the nineteenth-century housing intervention in industrialised countries was focused on dealing with overcrowding and unsanitary conditions. The problem with this form of intervention in countries like the UK was that there was no requirement to replace housing that was demolished and the result was therefore an increase in overcrowding.

The problem here was essentially one of political interest and the lack of majority representation. In many countries participation in politics was based on property ownership. But this meant that those who could vote or who were elected as representatives were the very owners of the properties targeted by reformers. More generally, there was a commonly held view that it was not the role of government to intervene in private property rights. There was therefore considerable resistance to intervention to improve housing conditions.

This was changed by three factors. First, the extension of the vote to all citizens, including the working classes, meant that poor housing became more of a political issue, and many political parties, particularly on the left, saw housing as a legitimate issue about which to campaign. The second factor that affected attitudes towards intervention was war. The two world wars in the twentieth century broke the resistance of many to government intervention in housing. After the First World War there was a clamour in several European countries to provide housing for those soldiers who had survived the carnage of the trenches, which saw the death of millions of their compatriots. In both world wars very little housing was built, which exacerbated pre-existing shortages, but the situation after the Second World War in Europe was exacerbated by the effect of bombing. Cities such as Berlin, Dresden, Rotterdam and Coventry were devastated by bombing and needed to be rebuilt after the war. Many countries in Europe undertook massive house-building programmes funded by government subsidies. Much of this housing was for rent and was specifically targeted at working-class households. The 20 years following the Second World War can be seen as the golden age of public or social housing across the developed world.

The third factor that influenced government intervention was the Great Depression of the late 1920s and 1930s. Many countries in Europe as well as the US undertook major public works programmes as a means of bolstering employment and economic activity generally. But also, particularly in the US, this period saw government support of owner occupation through the provision of mortgage guarantees. This view that government should take a more active role was widely accepted and given intellectual justification by Keynesian economics, and it continued after the Second World War.

Different countries used different means to subsidise housing, ranging from grants direct to landlords and permission to borrow supported by revenue subsidies, to mortgage guarantees for owner-occupiers and landlords. In addition to the provision of subsidies, governments have also attempted to regulate standards and to control rents of existing housing. Rent controls have been used across the world as a means of ensuring that housing remains affordable to those on a low income. However, this has had the consequence of reducing the incentive for private landlords to remain in the market, thus reducing the supply of rented housing.

The period after 1945 saw the growth in social housing, but also the growth of owner occupation. In some countries this was due to government support, but it was also due to increased affluence. Owner occupation became available to households lower down the income scale with the result that in many countries owner occupation had become the majority of tenures by the 1970s or 1980s. Governments were quick to see the political significance of owner occupation and have ensured that it remains the dominant tenure.

Running alongside this support for owner occupation was a general reduction in support for social housing. While some countries, such as the US, had never built much social housing, other countries such as Australia, New Zealand and the Netherlands ended object subsidies and shifted support to the demand side, providing direct personal subsidies to households. Many other countries took a more measured approach, but still there was a general trend away from the provision of direct subsidies and towards supplementing the income of low-income households, allowing them to compete in the market. Some countries such as the UK maintained a system of support for social housing, but through a form of mixed public and private funding. One effect of this was an increase in rent levels and so an increase in the cost of demand-side subsidies.

Much of the housing that households are living in now is older than they are. Indeed some of this housing is very sought after on aesthetic grounds or because of its location. But this also shows that housing is a long-lived asset, and this means that what we build now will have consequences for successive generations. This means that the standards that we build for one generation are important to the next. Likewise, the cost of housing affects future generations, particularly when housing is built through long-term borrowing, as is the case in the private sector but also increasingly in the social sector. We saw the consequences of borrowing, coupled with the continuing support for owner occupation, in 2008 when the collapse of parts of the US housing market helped precipitate a world-wide financial crisis. This is turn led to a world-wide depression in housing markets, with massive falls in house value in countries as varied as Ireland, Spain and the US. What this shows is that the seeds of problems were often planted many decades before and that the consequences of a policy can take rather a long time to have their full effect. Indeed, policies may have one effect in the short term, but a completely different one a generation later.

Discussion points

1   What attitude should government have to unplanned housing development?
2   Do we need to understand history to understand current developments?
3   What can we learn from other countries?

See also

Nearly all of the chapters are relevant here; see particularly:

Chapter 12 Property rights
Chapter 14 Social housing
Chapter 15 Private renting
Chapter 24 Rent
Chapter 27 Boom and bust
Chapter 28 Borrowing

Further reading

Harlow, M. (1995): *The People's Home?: Social Rented Housing in Europe and America*, Oxford, Blackwell.
Power, A. (1993): *From Hovels to High Rise: State Housing in Europe Since 1850*, London, Routledge.

4   The future

There are two qualities of housing that make it a complex problem. First, it is long lived. This is obviously a benefit in that housing lasts and can be used over a long period of time. But it also means that we are stuck with the consequences of past decisions, in terms of where existing housing is located, its size and the standards of amenity it has been built to. Second, housing is costly to produce and needs to be supported by long-term borrowing, government subsidies or both.

What compounds the complexity is that the type of housing that is required by a given population will change. The size of households can change over time; for example, it is common for household size to decline as societies become more affluent and when more women join the workforce. In addition, along with affluence comes greater longevity. This will mean that there are a greater number of smaller households, many of whom might have increasing support needs as they become frail with old age. A further issue is migration, both internal within a country, as people move for work or to improve their quality of life, and external, with economic migrants and asylum seekers entering from abroad. This means that the amount and type of housing we need will change over time, and if we are not careful we might end up with large amounts of the housing stock in the wrong place or of the wrong type. It is therefore necessary for societies to consider not just the needs of the present, but also those of the future.

There is then a need to understand the various demographic pressures on housing provision and plan accordingly to meet expected future needs. This means identifying trends in terms of changes in household size and migration. But inevitably this planning involves a degree of guesswork, there being many imponderable issues. It may not be possible to know in advance the level of immigration for 10 or 20 years' time, as migration is due to many economic and political factors that cannot necessarily be predicted. Yet, because migrants tend to be younger households, there will be consequences for the demand for family housing as well as for school places and access to health care.

So it may not be possible to accurately predict future housing needs. However, some trends are quite long term, such as whether a population is ageing, and so some planning is possible. Accordingly, governments will try to develop long-term plans in an attempt to deal with expected future housing need. This might involve planning targets and the requirement that local and regional government undertake regular needs assessment and plan locally.

This raises an important point. Housing is by definition local – it is stuck in the ground – and so the issue is whether the decisions on provision should be taken nationally or locally. There will be many regional differences in age of population, in terms of household size and in age profile. There will also be local differences in relative affluence and ethnicity. All these issues play a part in determining local needs. It might be argued, therefore, that local communities are the most suitable in determining local needs.

However, there may be particular local pressures that prevent needs being met. For example, a less affluent area might not be able to meet local needs. Or there might be local resistance to certain forms of provision. Thus, certain groups, who lack political power, such as recent migrants or Gypsies and Tra-vellers, might not be properly catered for. One way round this is for planning and provision to be determined nationally, with key decisions imposed on localities to ensure comprehensive provision and a more equitable distribution of resources. A further issue is that needs do not necessarily respect local boundaries. Households may work in one locality and live in another, being prepared to commute considerable distances in the process. This means that planners need to look beyond their immediate area.

The nature of planning is largely a matter of political culture, with some societies having a tradition of unitary planning whereas other political cultures are more devolved. But, whatever the culture, there will be the need to plan for the future and ensure that the housing that is being built is fit not just for the needs of today, but also for tomorrow.

Discussion points

1   What are the key demographic issues that determine housing need?
2   Who should determine what housing is provided?

See also

Chapter 3 The past
Chapter 7 Need
Chapter 33 Development
Chapter 34 Planning

Further reading

Dorling, D. (2008): *The Population of the UK*, London, Sage.

5 Ideology

Housing can be seen as a very practical matter. It is about bricks and mortar; it is about numbers and standards. Housing, then, is a matter of hard facts. But, while all these issues are important, they do not tell us everything about housing. It is not just a matter of numbers of dwellings, but also how we should live in a society. For example, do we have a right to live where we want, even if we choose to be in an expensive area and have a low income? And what level of responsibility do we have for others? When we talk about housing, we use terms such as 'public', 'social' and 'private', all of which come with a considerable amount of baggage. Housing provision raises important questions about the nature of society, and these cannot be answered simply by numbers. These questions connect with a moral and political vision of the type of society we wish to live in.

All of us have a set of beliefs that ground us and condition how we act. We can define this as an ideology: a set of beliefs that we accept and which we use to justify our actions. The term ideology is often used as a criticism. It is something that other people have, while we make decisions based on the facts. Yet, whether we admit it or not, we all have these bedrock beliefs and we use them to determine how we act.

Where it becomes more complex is when we suggest, as many Marxist commentators do, that there are dominant ideologies that are imposed on a society and which condition how that society is configured. The implication of this view is that somehow these commentators are immune to this ideology and are able to view the world differently from the rest of us. However, an alternative view is to see Marxism itself as ideology, just like liberalism and conservatism, and suggest that it too has its core beliefs that condition its critique of society.

Whether we take one view or another we need to be aware that the core beliefs that we have help to determine the very practical matters of housing. If we believe that individual households are competent and capable of making their own decisions, and if we believe that individuals alter their behaviour according to incentives, we might choose a form of housing that is based on private property ownership that allows households to choose where and when they move and how much they pay for their housing. Alternatively, if we believe that individuals are not capable of dealing with impersonal social forces such as inequality, poverty and social class, then we might conclude that the only agency that can provide good quality housing for all is a strong central government. So the types of provision that we see as necessary are guided by ideology. We might argue for social housing because we do not believe that some households can provide for themselves and that it is unfair for some to prosper while others suffer. Or we might promote owner occupation if we wish to encourage aspiration, personal responsibility and freedom from the state. The form of housing we take to be most important is determined, in part at least, by what we take to be a good society.

But, despite this, most people will still insist that they are not acting out of a particular ideology. Politicians and policymakers will state that they are acting on the basis of need and making rational decisions on the basis of the facts. Many individuals claim to be uninterested in politics and are only interested in taking care of themselves and their families. They do not see themselves as acting out of any ideological impulses, but rather just responding to what is in front of them and trying to do the best for themselves.

For those who recognise the importance of ideology, this response on the part of government and individuals will itself be seen an ideological response: self-interest and acting on the basis of rationality is itself ideological and masks what are particular interests that dominate society. This may be the case, but it also raises one of the problems of ideology and the arguments that surround it. Ideologies cannot be gainsaid. They are our core beliefs and there is nothing beneath them – this is as far down as we go – and accordingly we cannot readily discard them or replace them. For those who hold a particular set of beliefs, this is simply how the world is, and there is no means of separating them from their beliefs. It is for this very reason that we have to insist that everyone has an ideology, especially those who use the term to critique aspects of society that they do not approve of.

Think piece: The Right to Buy in the UK

The Right to Buy, introduced in the UK in 1981, can be considered one of the most significant housing policies of the last 50 years, but it is also perhaps the most controversial. The policy allowed sitting tenants in social housing to buy their current dwelling at a discount. There was an initial qualifying period of three years and a maximum discount of 50 per cent for houses and 70 per cent for flats. The effect of the policy was to reduce the stock of social housing in the UK by 2.5 million dwellings, equivalent to more than 40 per cent of the entire stock.

The rhetoric that supported the policy emphasised personal responsibility and independence from the state. Households would gain control over their dwelling and have an asset that they could pass on to their children. This was very much linked to the new agenda developed by Margaret Thatcher when she became leader of the Conservative party in 1975. Once the Conservatives took power in 1979 they were able to implement this new agenda that included a major promotion of owner occupation.

Much of the discussion around this promotion on owner occupation, including the Right to Buy, concerned the link between property ownership and voting behaviour. The belief of the Conservatives was that encouraging owner occupation was a clear vote winner and would cement the apparent trend of working-class households to vote Conservative. Certainly the land-slide Conservative victory in 1983 appeared to confirm this, as their main opponent, the Labour Party led by Michael Foot, was committed to ending the Right to Buy.

Clearly, however, the link was either non-existent or temporary, in that the Conservatives were voted out of office in 1997. But, by this time, the Labour Party, as well as the other parties, had become reconciled to the Right to Buy. Between 1997 and 2010 there were some restrictions placed on pro-spective buyers, lengthening the qualifying period and lowering discounts. But there was no attempt, with the exception of Scotland, to repeal the policy. Indeed, following the 2010 election the Conservative-led Coalition reinvigorated the Right to Buy, increasing discounts and reducing the qualifying period.

What this suggests is that the Right to Buy has achieved such a status amongst politicians that it is unrepealable. The rhetoric has become so ingrained that no politician can contemplate getting rid of the policy. This shows the power of ideas in politics and the belief that there are certain policies that are unassailable because of general public approval.

But the effect of the sale of so much social housing has been con-siderable. Not only did it reduce over time the available relets, it also reduced the rental income of social landlords. The properties sold were disproportionately made up of family housing, which reduced the ability of landlords to help homeless families and left landlords with a higher level of unpopular and hard-to-let stock. The effect of the discounts, and the constraints on the reuse of capital receipts, meant that dwellings sold could not be replaced. Despite this, however, it is perhaps more likely that the Right to Buy will be extended in the future rather than abolished.

References

Green, E. (2006): *Thatcher*, London, Hodder Arnold.
Jones, C. and Murie, M. (2006): *The Right to Buy: Analysis and Evaluation of a Housing Policy*, Oxford, Blackwell.
King, P. (2010a): *Housing Policy Transformed: The Right to Buy and the Desire to Own*, Bristol, Policy Press.

Discussion points

1   What are the links between ideology and tenure?
2   Is there a dominant ideology?
3   What are your core beliefs and do they matter?

See also

Chapter 11 Owner occupation
Chapter 12 Property rights
Chapter 13 Desire
Chapter 14 Social housing

Further reading

Adams, I. (1993): *Political Ideology Today*, Manchester, Manchester University Press.

Part 2

Concepts

6 Social justice

Over the last 100 years many politicians and commentators have used the argument of social justice to justify the welfare state and government provision more generally. Interestingly, the concept has usually been associated with the political left and has been associated with justifications for social housing provision. However, in recent years, right-of-centre politicians in the UK and US in particular have sought to use the concept to justify their policies. This has caused some controversy, as much on the right as the left. Over the last 30 years there have been a number of critiques of social justice from the right that are relevant to housing and welfare more generally. So it will be useful to consider the concept of social justice in some detail to look at what it means and how it has been and is used.

The best way to tackle the concept of social justice is to decouple the two words and look at them separately. So we shall begin by exploring the concept of justice and then add the concept of the social back in later. David Miller (1976) states that a simple definition of justice would be 'to each his due' (p. 20). But he goes on to elaborate on this: 'The just state of affairs is that in which each individual has exactly those benefits and burdens which are due to him by virtue of his personal characteristics and circumstances' (p. 20). Justice therefore involves some calculation over what is proper and proportionate. We need to determine what is a proportionate response to a person's actions based on the relevant attributes that the person possesses. So, in the sense of a criminal action we would need to look at the nature of the offence, but also to see whether there were any mitigating circumstances. For example, was it a case of cold-blooded murder or was the murderer provoked beyond endurance by domestic violence on the part of her partner?

The most common way in which we see justice is in the legal sense. This is concerned with punishment and compensation through the creation of a public set of rules (or what we call laws). Principles of justice here relate to the conditions under which punishment may be administered and restitution made, but also to the procedures for applying the law. Thus one can talk of both a just punishment and a just process for arriving at a settlement.

So, justice is a concern for the distribution of burdens and benefits based on certain relevant characteristics that individuals or groups might have. This allows Miller to extend the concept to the social sphere. Accordingly, he defines social justice as concerning 'the distribution of benefits and burdens throughout a society, as it results from the major social institutions – property systems, public organisations, etc.' (p. 22). Thus social justice deals with wages and profits, the protection of rights, the allocation of education, health care, housing and so on. It is a concern for the allocation of resources throughout a society on the basis of some normative principles of what each individual is due.

Undoubtedly the most important theory of justice developed over the last 50 years is that of John Rawls in his work *A Theory of Justice* (1971). Rawls's theory is a social contract theory, in that it presupposes under what conditions individuals would come together to form a society that would protect their interests and fulfil their needs. Rawls poses the question of what principles would a group of rational individuals derive if they sat down to create a society that maximised both freedom and equality. He argues that such a deliberation would lead to the derivation of two principles. First, there is the principle of autonomy, whereby society allows the greatest possible liberty for the individual, compatible with a similar degree of liberty for all. Second, Rawls proposes what he calls the Difference Principle, which deals with equality in society. Rawls suggests that inequalities and major differences in income and wealth are justified only insofar as the disparities of wealth contribute to the greater benefit of the least well off. Rawls recognises that free and rational individuals respond to incentives and self-interest. This means that a certain degree of inequality may be necessary to provide for certain necessary occupations such as doctors and teachers. Why, we might argue, would someone train to be a surgeon for over a decade and then only be paid the same as a hospital porter? Rawls acknowledges that, if we want highly skilled professionals, and entrepreneurs who are pre-pared to take risks but who might create many jobs for others, then we need a degree of inequality. Yet this inequality should be limited to the level that benefits the least well off. Inequality therefore has to be justified by the benefits that accrue to society particularly in the form of provision for those at the bottom of society.

Rawls would have us believe that these two principles are not arbitrary, but rather would derive from a social contract made by rational individuals seeking to protect themselves. What he asks us to do is to imagine that individuals within a society are in what he terms the 'original position'. In the original position individuals are placed behind a 'veil of ignorance', whereby they are aware of the general facts about human nature – incentives, self-interest and so on – but know nothing about their own birth, wealth or talents. Rawls argues that these are all morally arbitrary and so should be discounted in any deliberation on outcomes. This means, though, that each individual who is party to the con-tract does not know whether they would do well or badly from the agreed social arrangements. Accordingly, Rawls argues that any rational individual who justifiably fears that they may be worst off will seek principles which both maximise the possibility of freedom, allowing them to better themselves, *and* limit inequality to the level that would benefit them.

Thus, appreciating human nature, they will concur that incentives need to be provided to ensure a sufficient supply of surgeons, teachers, entrepreneurs, etc. Thus they will opt for equality, except where inequality benefits them. These two principles can then be used to distribute the 'primary goods', such as income and wealth, rights and liberties, power, self-respect and equality of opportunity, which every rational individual is presumed to want. This connection of primary goods and social justice allows us to make a connection with more concrete policy objectives such as the allocation of housing.

I have suggested that many on the right have been critical of social justice. These criticisms are important in that these critics had some influence on many governments in the 1980s and 1990s. Thinkers such as Robert Nozick and Friedrich Hayek informed the reforms of politicians such as Ronald Reagan in the US and Margaret Thatcher in the UK.

The first critic of social justice was a colleague of Rawls at Harvard University, Robert Nozick, who published a work entitled *Anarchy, State and Utopia* in 1974. This work is concerned with critiquing Rawls but also provides an outline of what he terms the entitlement theory of distributive justice. We can understand his critique of Rawls better if first we discuss his entitlement theory. Nozick's starting point is to state that:

> There is no *central* distribution, no person or group entitled to control all the resources, jointly deciding how they are to be doled out. What each person gets, he gets from others who give to him in exchange for something, or as a gift. In a free society, diverse persons control different resources, and new holdings arise out of the voluntary exchanges and actions of persons .... The total result is the product of many individual decisions which the different individuals are entitled to make.
>
> (1974, pp. 149–150)

From this starting point, Nozick develops the idea of a just distribution based on entitlement to property (or, as he terms it, holdings). He suggests that this consists of three principles. The first principle derives out of the *original acquisition of holding*, or 'the appropriation of unheld things' (p. 150). Thus, 'A person who acquires a holding in accordance with the principle of justice in acquisition is entitled to that holding' (p. 151).

Second, which in practice would be the main form, there is the *transfer of holdings*, which deals with voluntary exchanges and gifts. Thus, 'A person who acquires a holding in accordance with the principle of justice in transfer, from someone else entitled to the holding, is entitled to the holding' (p. 151). No one is entitled to a holding, according to Nozick, except by repeated applications of these two principles.

However, some holdings have derived from fraud, enslavement or other illegitimate action. There is therefore the need for a third principle, which Nozick refers to as 'the rectification of injustice in holdings' (p. 152). This concerns dealing with the question: 'If past injustice has shaped present holdings in various ways, some identifiable, some not, what now, if anything, ought to be done to rectify these injustices?' (p. 152). These three principles determine whether an individual's holdings are just and 'If each person's holdings are just, the total set (distribution) of holdings is just' (p. 153).

Nozick is offering here a historical account rather than one based on outcomes like Rawls. One derives the justice of a distribution from how that distribution has been arrived at. So what matters is the process of distribution and not the outcome.

According to Nozick, a key distinction between his theory of justice and others is that not only is his theory historical, but it also does not specify a particular pattern or end result. In this way he connects up justice and individual freedom. A patterned principle is one that prescribes the form of distribution or prejudges outcomes within a society. Thus it is where a particular distribution (an end result) is seen as just. However, for Nozick, this situation cannot be just as it involves the continual interference with individual liberty to maintain this pattern of distribution. Accordingly, Nozick sees Rawls's theory of distributive justice as being an end result and therefore infringing on individual freedom.

Nozick's view might be philosophically sophisticated, but his views have not been as influential on the practical level as those of Friedrich Hayek. Like Nozick, Hayek's position is one of the process of justice rather than outcomes, but his is also more directly critical of the whole concept of social justice. Hayek sees the concept of social justice as a mirage. It is based on a fallacious coupling of what ought to be two distinct entities, justice and the idea of the social. Hayek argues that social justice is an abuse of the word 'justice', which 'threatens to destroy the conception of law which made it the safeguard of individual freedom' (1982, vol. 2, p. 62). The problem with the term 'social' is that it encourages the placing of the artificial values of a centrally controlled society above individual morality on the assumption that the aims of society are superior to those of individuals.

A proper use of the concept of justice would be a rule that could be generally applied equally to all citizens. For Hayek, justice can only relate to the treatment accorded to individuals, and therefore enforced distribution (i.e. that which does not have their express consent) is by definition unjust, as it treats individuals differently. Hayek sees that this may lead to outcomes that are unequal or unfair, but that these outcomes are not unjust if they have derived from rules which affect everyone equally.

The importance of these critics is that they gave ammunition to right-wing politicians to dismiss the very idea of social justice, and for a time this was successful, with concepts such as choice and aspiration being at the centre of public policy. What mattered, as Hayek argued, were the interests of individuals and not society. In terms of housing policy this manifested itself in the promotion of owner occupation and the greater use of subject subsidies.

However, since the start of the twenty-first century there has been something of a renaissance for the concept of social justice. But, instead of this coming from the left it has been promoted as much by the right. We can see this concern in the policies to promote owner occupation for minority house-holds of George W. Bush in the US, but also in the rhetoric and policy of the Coalition government in the UK, particularly with regard to welfare reform. However, there is a markedly different emphasis in this conservative form of social justice. Instead of increasing welfare expenditure and entitlements, social justice can only be attained by altering incentive structures and the culture of dependency that maintains communities in poverty.

But the similarity with older versions of social justice is that change can only be achieved through an active government rather than the free market approaches advocated by Nozick and Hayek. It also assumes that government can shape society rather than leaving it merely to individual morality and decision-making. This may not convince those who wish to adhere to a more traditional view of social justice, but it does show the enduring appeal of the concept.

Discussion points

1   Can we justify social provision using social justice?
2   What is the main objection to the concept of social justice?
3   Why do you think some conservatives now find the concept of social justice appealing?

See also

Chapter 14 Social housing
Chapter 16 Welfare
Chapter 17 Poverty
Chapter 18 Fairness
Chapter 19 Inequality

Further reading

Hayek, F. (1982): *Law, Legislation and Liberty*, London, Routledge and Kegan Paul.
Miller, D. (1976): *Social Justice*, Oxford, Oxford University Press.
Nozick, R. (1974): *Anarchy, State and Utopia*, Oxford, Blackwell.
Rawls, J. (1971): *A Theory of Justice*, Oxford, Oxford University Press.

7 Need

The concept of need is central to any discussion of housing, and particularly to how it is allocated and who gets it. Hence determinations of need have been used to allocate social housing, but also to decide on what new housing should be built and where it should be located. The importance of need as an argument is that it is directly opposite to the idea of market provision, which is based on the ability to pay. Need is where social criteria dominate, rather than individual choice and commercial considerations.

The clear implication is that, when need is involved, it is not proper for a landlord to make a profit or to put their private gain above or even alongside the interests of tenants. Need, it is suggested, overrides commercial considerations. For this to happen, of course, housing provision would often have to be organised outside markets and according to a set of principles which emphasise the tenants' interests.

But how is it possible to operate outside of markets? Clearly it can only be done if it is possible to define housing need and to identify those who are to be helped. A particularly useful definition of need is that provided by Ray Robinson (1979). He defines housing need as:

> The quantity of housing that is required to provide accommodation of an agreed minimum standard and above for a population given its size, household composition, age distribution, etc. *without* taking into account the individual household's ability to pay for the housing assigned to it.
>
> (pp. 55–56)

This type of definition is sometimes referred to as geographical need, in that it seeks to identify the housing requirements for a particular population, say, within a local authority or municipality. But, of course, this definition could also be used to apply to a region or even a national population.

There are a number of important elements to this type of need. First, Robinson points to the fact that housing need requires the establishment of a particular standard of housing provision. This may be defined in statute in terms of fitness and habitability or it might be stated in policy terms. Second, this definition of housing need excludes the ability to pay as a criterion. Housing need should be determined by objective conditions, such as household composition and the standard of the housing stock, and not according to income. The idea behind this is that all households should be able to gain access to housing of a certain standard and that this applies regardless of income.

Third, and perhaps most significant for our discussion here, need is defined externally. It needs to be assessed by experts from outside the particular population on the basis of formally established criteria. It is not the individual households within the population who determine their needs, but rather they are deemed to be in need by virtue of how they measure up against the particular standard that has been agreed for that population of which they are a part. It is not for individual households to determine whether they are in need, but for experts to make that judgement.

Indeed this definition is a rather impersonal one, in that it tends to look at whole populations rather than their component parts. This is a valid exercise, in that it may show the scale of a problem and allow a local housing organisation to make an informed bid for government funding. However, it does not cover the issue of need completely. We can see this when we consider the manner in which social housing is allocated. In order to do so, a population has to be disaggregated to allow for the allocation to be made. The landlord needs to be able to differentiate between rival applicants to determine who should be allocated a vacant dwelling. This, of course, could be done on the basis of 'first come, first served' or according to who agrees to pay the highest amount of rent. But, if one is seeking to allocate to the most deserving or most vulnerable, the landlord has to be capable of differentiating between the needs of different households. Likewise, in means-tested benefits systems there has to be some way of targeting the benefits for those who need them most.

What is needed, therefore, is a means of differentiating between the needs of individuals. The type of need defined by Robinson will merely identify the number and type of properties needed in an area. However, what is required is some means of deciding which households within that area should be allocated these dwellings.

Jonathan Bradshaw (1972) has distinguished between four types of need that can be used to separate households. However, before discussing them, it is interesting to note that Bradshaw refers to these needs not as 'individual' but as 'social'. Social need is what society as a whole identifies as a problem or a lack, which it seeks to remedy through the provision of a social service, such as social housing. So the need is social in the sense of identifying a problem which society wishes to see eradicated. However, the identification of these social needs can be used to differentiate *between* different households and so use need as a means of allocating social housing to individuals.

The first category that Bradshaw identifies is normative need. This most closely relates to geographical need and can be seen as what some expert or authority defines as need in a given situation; indeed it can be used to identify the needs of a population, as well as to distinguish between its members. In essence it suggests that we identify some acceptable norm or standard to which everyone should have access; hence the similarity to Robinson's definition of defining housing need according to a minimum standard. It is a normative standard in that the actual level of provision is dependent on the specific time and place, rather than being an absolute. If someone falls below that standard then they are in need and should be helped.

Second, Bradshaw identifies felt need, which refers to an individual's own assessment of their requirements. This might be assessed by surveys, questionnaires or interviews, and can therefore be highly subjective. In particular, it can often be difficult to separate out a want from a need. However, such a notion of need is important in that it can provide data on consumer satisfaction and on the perceptions that consumers have of a product or an agency.

Third, we can identify expressed need, which is where the felt need is acted upon. This is shown by our purchasing behaviour, or by what economists call effective demand. But this too does not distinguish between a real need and a want: just because we have money to spend on something, it does not mean we need it. However, it can be seen as a means of determining what policies and practices a social landlord undertakes and what sort of dwellings it builds. We can perhaps see choice-based lettings systems as an example of this type of need, in that applicants actually have to bid for specific properties in particular locations. This can be seen as a more accurate guide to what people need (or want) than answers to hypothetical questions in a survey.

Finally, Bradshaw defines comparative need. This is when a comparison is made with those who are already in receipt of a service. It is when we compare those who are well housed with those in similar circumstances who are not well housed. The latter are then said to be in comparative need. This type of need emphasises equal treatment and fairness.

These are indeed useful categories, but what we still have not come to is a precise definition of need. All the discussion above has presupposed is that we know what a need is. Yet, as the discussion on Bradshaw's four types of need shows, it can be difficult to separate a need from a want. But, if we are to take the concept of need seriously, we should have a means of separating needs from wants; otherwise we cannot justify prioritising either a particular individual or group, or the necessity for certain types of dwelling rather than others.

So we should seek out a precise definition of a need that helps us to distinguish between imperatives and aspirations. As a starting point we can distinguish between needs and wants. We might suggest that needs are things imposed upon us regardless of our conscious will, whilst wants are things we choose for ourselves as a means self-expression. Wants may not be things we absolutely have to have, but they relate to our perception of ourselves, our aspirations and the status we seek. It is not therefore a matter of whether we have a house or not, but rather what type of house, where it is, what the neighbours think and what the house says about us. What is immediately clear is that Robinson's definition of housing need could equally refer to wants. What would determine the matter would be the actual minimum standard that is set. Modern standards do not just relate to keeping us warm, dry and safe from intrusion, but go much further to include a high degree of comfort and leisure.

We might argue that wants can be met through a market, whereas needs may be better met by government action. Thus wants are related to choice, but needs are not. The question, of course, is where one draws the line between a want and a need. Supporters of markets might state that there is actually no such entity as need. What are referred to as needs are in fact individual desires and preferences. In other words, they are subjective wants that allow for no universal statements to be constructed about the general human condition. According to this view, free marketeers do not accept any objective notion of need as the basis for decisions about what society requires. Instead they would seek a much greater reliance on markets. But this view can be opposed by those who suggest that need can form an empirical and objective basis for social policy, whilst admitting that there is still a normative element to the formation of any concept of need.

An important example of this is provided in the work of Len Doyal and Ian Gough (1991), who see the concept of need as offering the potential for a universal statement on social provision. They argue that there are two basic human needs that are required to ensure the 'avoidance of serious harm' (p. 50) and that these apply regardless of time and place. The first basic need is personal autonomy, which they define as 'the ability to make informed choices about what should be done and how to go about doing it' (p. 53). The second basic need is physical survival and health, which is described as the ability to carry out necessary actions. The loss of either autonomy or health would entail disablement and an inability to lead anything near to a normal life, and this would apply regardless of time, culture or place. According to Doyal and Gough, these two basic needs are therefore universal. But, in addition to these two basic needs, they suggest that there are a number of universal satisfiers, such as food, water, security and housing. These are derivative of the basic needs, in that they are required to maintain our autonomy and health and so ensure that we avoid serious harm. However, whilst they are universal, the actual level required for satisfaction will differ according to time, place and culture: we all need housing to protect our autonomy and health, but what this amounts to in terms of space standards, building type, size, etc., is relative to a particular culture.

What is clear from Doyal and Gough's discussion is that a need is an imperative. If we are without this thing we need, the consequences may well be severe. It is not merely a case that we are disappointed that our aspirations have not been met and that we have had to choose something else in its place. If we do not have our needs met, we would be at risk of serious harm and danger to our lives. This means that our needs require immediate attention. We cannot ignore them or put off dealing with them for very long.

Doyal and Gough's formulation of universal needs carries with it the probability that individuals may be unaware of what some of their needs are. Stated differently, we can suggest that it is not necessary to be aware of our needs in order to have them. Such needs will therefore need to be determined, to an extent at least, externally. This does suggest, then, that needs can only be fully defined externally to the individual or group concerned. It also assumes that needs are capable of objective definition, measurement and assessment and can therefore form a legitimate basis for government action.

This view is confirmed by a more philosophical discussion undertaken by James Griffin (1986). He concurs that wants and desires must be intentional states in that we must be aware of the condition. We purposefully want or desire something, and so Griffin sees that a want or desire is tied to our perception of an object. However, need is not an intentional verb. It does not have to be related to the perception of an object, nor to our particular experiences. We need something only because it is necessary, not because we crave or desire it. According to Griffin, we do not have to know that we have a need for it to exist, whilst we must actively want or desire something. Needs exist regardless of our consciousness of them.

We might find Griffin's argument initially quite strange: how can we have a need and not know anything about it? However, a small child or an elderly person suffering from dementia would have no conception of the importance of avoiding serious harm or appreciate the significance of personal autonomy. A small child might not know that a busy road is dangerous and that they need to take care, and nor could they read the word 'poison' written in big letters on the bottle accidentally left within reach. But it is not just the very young and the old who may be unaware of their needs. We might not be aware that we are suffering from cancer until it is too late, and this is simply because we are unaware of the symptoms. Similarly, we might ignore some discomfort or misinterpret heart problems as heartburn. But our ignorance of the true state of our health is irrelevant to our situation: the cancer or heart problem is there and needs treating, whether we currently know about it or not.

This is perhaps the most significant point when considering the assessment and fulfilment of need. If individuals are unaware of their needs, they must either be made aware or have the need met for them. Hence some societies have compulsory social insurance schemes to force people to save for the time when they are ill or elderly. Other states provide goods and services like health care centrally, regardless of the ability to pay, so that there are no costs involved and no excuse not to deal with the discomfort.

However, this does not mean that we have no choice with regard to need. Griffin suggests that there are two sorts of needs. First, there are *basic needs*, which we all have by virtue of being human. We can see these as being similar to those defined by Doyal and Gough. But, second, Griffin states that we have what he terms *instrumental needs*, which occur because of the particular ends we choose. So, for example, we may choose to have a child, and this can quite properly be seen to be a voluntary action. However, once a woman is pregnant, she now has certain needs as a result and, once the child is born, the household has additional needs in terms of space, income and health care. Like having a child, many of the choices that we make are consequential, in that once we have made the choice we cannot rescind it, or at least not without equally serious consequences (we only need to consider the moral and psychological dilemma of abortion to see this).

This suggests that many of the needs that we now have have arisen because of the choices we made earlier, be it the relationships we have had, the jobs we have taken or the commitments we have made to others. This has important consequences for policymaking. As an example, we might say that being without a dwelling carries with it the potential for serious harm. Therefore, we would hope that the household would be helped as a priority. But does it not matter how the household became homeless? What if it was due entirely to their own actions: they perhaps neglected to pay their rent and so were evicted? Should this affect the way in which we deal with them? In policy terms, there is a distinction between homelessness legislation in England, which retains a test of intentionality precisely to deal with deliberate and avoidable acts and omissions, and Scotland, where the intentionality clause has been abolished on the basis that what matters is dealing with homelessness and not apportioning blame. So English and Scottish legislators persist with different approaches to instrumental needs.

What this means is that it is never possible to separate need from choice completely. Individuals have needs because of the choices they have made, and these needs do not go away merely because we see them as less serious than basic needs. This is particularly the case in developed economies where most people live well above a basic subsistence level. Many of the state's resources are not used to maintain basic needs but go well beyond this. Hospitals provide care to a high level, and in countries like the UK it is possible to receive fertility treatment and some forms of cosmetic surgery. However, most UK citizens would not see this as excessive or improper. It is rather the case that we are capable of successfully intervening and therefore we should. Indeed, debates concerning health care tend to be about the lack of sufficient resources to fund certain identified needs, either because drugs are deemed too expensive or because different policies pertain in different parts of the country. The issue here is that new needs have effectively been created by technological advances and medical breakthroughs.

What is significant about this development, however, is that UK hospitals do not provide fertility treatment *instead* of basic care but *as well as*. Likewise, when social landlords provide high quality housing, they cannot help but meet the very basic standards. Luxuries, as it were, sit on top of basic requirements, and so living in a mansion would see our basic needs fulfilled, as well as matching up with our aspirations.

This suggests that a discussion on need alone would be rather limiting. Of course, we want basic needs to be met, and this may serve as a sound justification for housing subsidies and state provision. However, it is not enough in itself. We want and expect to go beyond basic needs and would not be happy to return to a subsistence existence. We want and expect to be able to choose and to have a civilised life that is determined by the current conditions in which we live. In other words, we want to be able to make choices and decide for ourselves. We are simply not happy allowing others to take the major decisions for us.

Discussion points

1   Does it matter that needs are normally determined externally to the individual or group in question?
2   Is need the best means of allocating scarce resources?
3   Can need and choice ever be properly separated?

See also

Chapter 8 Choice
Chapter 10 Responsibility
Chapter 14 Social housing
Chapter 17 Poverty

Further reading

Bradshaw, J. (1972): 'The Taxonomy of Social Need', in McLachlan, G. (Ed.): *Problems and Progress in Medical Care*, 7th series, Buckingham, Open University Press.
Doyal, L. and Gough, I. (1991): *A Theory of Human Need*, Basingstoke, Macmillan.
Griffin, J. (1986): *Well-Being: Its Meaning, Measurement and Moral Importance*, Oxford, Clarendon.
King, P. (2003): *A Social Philosophy of Housing*, Aldershot, Ashgate.
Robinson, R. (1979): *Housing Economics and Public Policy*, Basingstoke, Macmillan, pp. 55–56.

8 Choice

In some housing tenures, choice is the norm. Owner-occupiers are able to decide when they move and what they move to. Of course, this is limited by income and circumstances, but there is considerable latitude for the average owner. The same applies to some private renters, albeit to a lesser extent in that they are more constrained by their landlord, who retains certain rights over the property, often including how long they are prepared to let their tenant remain in their dwelling. However, for those dependent on state subsidies, the level of choice that a household can have is significantly reduced. These households are dependent on the policies of landlords, but also the nature of the subsidy system and specific priorities determined by legislation. Indeed, we might wonder why choice is even relevant for households who are deemed to be vulnerable or in priority need. What matters more is making sure that they are properly and safely rehoused, and the niceties of choosing ought not to come into it. Some might even argue that people who have got themselves into a mess might not be very good at making choices in the first place.

This latter view, however, can be seen as rather patronising and even unacceptable in a world that is based on individuals making choices over many areas of their lives. People can choose what to wear, what to eat, whom to associate with, what to watch and who to vote for, so why cannot they choose where they live? If individuals are consumers when it comes to shopping for clothes, why not shopping for housing? Indeed there has been a general shift towards more consumer-oriented structures across the developed world. Many countries have shifted towards demand-side subsidies paid to households rather than housing organisations. In principle this allows the household to determine what housing they live in rather than waiting passively for a dwelling to be allocated to them according to someone else's priorities.

Not only does this shift in policy seek to empower households and alter the relationship between landlord and tenant, it also implies a different function for housing subsidies and government involvement in housing. Subsidies paid to individuals assume that 'the housing problem' is a lack of income and access to housing, rather than a shortage of housing itself. Government intervention is geared towards ensuring that households can compete in markets rather than encouraging new development. So, shifting the emphasis away from need and towards choice is important in showing a change in the direction of policy that favours households instead of landlords. It suggests that we see who owns the housing as being less important than it was in the past. Instead of seeing that the best means to help those in need is to build houses for them, we now see that it is better to increase their income to allow them to find suitable housing for themselves. It might also indicate that the role of experts who can make objective judgements on the needs of others is being questioned. Instead we should place more emphasis on the ability of individuals to make decisions for themselves.

But is choice always a good thing? What if households, who are now expected to make decisions themselves, prove to be incapable? What if they make a bad decision and they and their children end up homeless? Is this a price worth paying for allowing people the freedom to choose? And, just who is capable of deciding whether choice is a good thing or not?

One of the problems when discussing choice in housing is that there has been all too little reflection by policymakers on what is meant by the term. Like need, it is one of those concepts that we use all the time, and we assume that we know what it means. Often it is seen as a quality which individuals carry within them, which can be released by the right sort of policy mechanism: individuals 'have choice', and all that is needed is the opportunity to exercise it.

However, this is too simplistic: the concept of choice is actually a rather complex one that carries with it certain moral implications for both consumers and organisations. Indeed, the allocation of housing is itself a moral issue: making a decision about allocating a government-funded dwelling to one household rather than another is not a value-neutral process. We need to have reasons for choosing one household and not the other. It may be because the household chosen is more deserving, or in greater need, but we are still choosing one person instead of another. Likewise, when we decide to provide subsidies to households rather than landlords, we are making a judgement on the competence of individuals to decide for themselves, as well as implicitly commenting on the ability of landlords to act properly on their tenants' behalf.

Choice, in a moral sense, relates to notions of autonomy, liberty and responsibility. In this sense, a choice-based policy should be one that alters the power relations between landlord and tenant/applicant in favour of the latter. In turn, a greater burden is placed on applicants and tenants in terms of bearing the responsibility for their decisions. Hence it is entirely proper to connect choice with empowerment: having a choice should mean that one has more control over one's immediate affairs.

Put simply, to have a choice or to choose suggests that we are able to select from alternatives, even if the alternative is an either/or between two less-than-perfect solutions. It further implies that we are able to make a preference and thus distinguish between entities, and that we are able to proffer reasons for the choices we make. Choice is deemed to be a capability that individuals and households have, whereby they can materially affect their situation through the decisions they take. It is the point at which individuals take control over the decisions affecting them.

This raises what can be considered the most significant question in the debate over choice, namely, what is the level of knowledge that is presumed necessary in order to facilitate effective decision-making? As the Norwegian political scientist Jon Elster (1986) has shown, one of the essential prerequisites for rational choice is information. We need access to accurate and correct information before we can come to a considered decision.

According to Elster, in order for an action of ours to be rational, it has to relate to our desires, beliefs and information. He argues that an understanding of rationality involves the interplay of these three levels of desires, beliefs and information sets. First, for an action to be rational it has to be the best means of satisfying our desires, given our current beliefs. There should be no better way of satisfying our desires; otherwise the action is not rational. Second, Elster argues that the beliefs themselves should be rational, meaning that the action should be based on concrete information, which again includes our beliefs. However, Elster acknowledges that it might be the case that we are not aware of the full set of opportunities open to us. Thus, we may be able to do more than they believe we can. Equally we might overestimate our opportunities and overstretch.

Connecting rationality to belief emphasises its subjective nature. Rational choice involves making some subjective means of ranking alternatives. This means that we can fail or be mistaken in the chosen action, without that action being irrational. According to Elster, an action is rational if we have no reason in hindsight to think that we should have acted differently. According to this view, a drug addict may be said to be acting rationally if he or she is a person who subjectively discounts the future very heavily, so that the only thing that matters is the next fix and the money to pay for it.

The third element of Elster's discussion of rational choice is the need for information. Elster suggests that there is a balance to be made in information-gathering since we cannot make a rational decision without investing time and effort in doing so. Yet it might also be the case that gathering too much information is dangerous. We would hope that a doctor does not wait too long to make a diagnosis, in case the patient dies. We should therefore seek some medium between considered diagnosis and decisive action, although Elster acknowledges that it might be difficult to locate where this point is. This also reminds us that making choices takes time and this might not be available if the needs of the patient are very great.

What makes this balancing act more difficult, of course, is the asymmetrical nature of information. Doctors know a lot more about medicine than we do, and so how do we judge when it is the right time for them to arrive at a correct diagnosis? Likewise, private landlords may have more knowledge of market conditions and rent levels within a district, which would provide them with a significant advantage over applicants. The same asymmetrical relationship would apply in the social sector, where the necessary information required by the applicant is filtered through the landlord's bureaucratic structures. This means that the costs of information-gathering are often determined by landlords.

Putting the three elements together leads Elster to state that rational choice is the principle that people make the most out of what they have. This definition of rational choice is an important one. It recognises that rationality is a subjective notion and thus depends on desires and beliefs and information available to the decision-maker. In doing so it recognises that individual decision-makers do not start from positions of equality: 'what they have' will differ between individuals. This shows that the key issue is that of the resources available to the decision-maker.

Elster's discussion shows that choice is always bounded by constraints. He sees these constraints as relating to the incentive structure within which the decision-maker operates and the limited information that they may have. What this means in practice is that we often find ourselves in situations where we have to choose. Indeed, we may find we have no choice but to choose! Choice is often triggered by some event and so it is not a continuous event. Choice is therefore contingent on certain key events, such as a job change, retirement, or relationship breakdown. Households only see the need to choose as a necessary reaction to a trigger event. There is thus an important instrumental quality to choice: we choose because of the situation we are in and because we have to react. Furthermore, we choose *between* specific entities rather than having an inner quality that we might exercise as it suits us. We do not choose for its own sake but to achieve a particular effect.

A further issue with choice, that is particularly relevant when we are talking about housing that might be allocated by bureaucratic means, is who gets to choose first, and how do we decide this? Clearly, the person who gets to choose first has the advantage of being able to pick from all that is available. Those further down the queue will have fewer options but can still be said to have a choice. But, for everyone to have a choice, there needs to be an absolute surplus of resources. This raises three issues. First, this situation implies that it is not just a matter of whether we can choose, but what we can choose and when. Do we have to wait for others to choose first and, if we do, how can this be justified? Who decides on who chooses first and, more fundamentally, who determines whether we can choose at all? This latter point matters because, as we shall see below, sometimes people make bad choices and the consequences for them and their families might be considerable.

The second issue is that, if we need to have a surplus for all to have a choice, then it is actually wasteful of scarce resources. Would it not be better, we might argue, for the state to plan its resources according to actual need rather than allowing all individuals to have a choice?

Third, we might wonder whether we are always the best person to decide on what is good for us. We have already suggested that we might have to rely on experts in certain fields such as medicine. But, more generally, on what basis can we say that we know best and, in particular, do we really know where the limits of our choice-making capacities are? The only sure way of knowing is to allow people to make choices until they fail, but do we really want this to happen?

Choice is meant to allow individuals to influence their own situation materially. The idea of choice-based policies is to allow individuals to make their lives better. The problem is that choice over public services cannot be made autonomously, but is in the gift of a bureaucracy determined to control it. Landlords control the resources and the access to them, and this means that it is landlords who determine the level of choice that applicants and tenants actually have. Likewise, those on low incomes need subsidies determined and controlled by government before they can make choices. Therefore, we can suggest that the essence of choice is who has control over resources.

If we are to make the most of what we have, it therefore matters what we have and how much we are able to make of it. One way of describing this is as effective choice. This can be defined as the ability of individuals to control their environment and gain access to the resources that translate choice into empowerment. Effective choice exists, therefore, when we are capable of controlling our environment, as opposed to having a merely abstract choice. Individuals with effective choice could be said to be empowered and thus capable of making the most of what they have.

We can define three principles for housing processes based on effective choice. The first principle involves the limitation of the role and scale of government activity in housing. Central government does not need to set down distinct lines for action, but rather it should merely set limits or parameters within which agencies and individuals can operate. Thus government's role should be restricted to setting limits to action which allow for the maximum opportunity for individual fulfilment. Individuals can only choose if they are given the space to do so, and this has implications for financial mechanisms and who controls them.

Second, the control of the housing process should be local and in the hands of those who use the outcomes of the process. The smaller the scale, the better the outcomes can be manipulated by the users. This presupposes that choice-based systems should be 'bottom-up' and not determined by central government prescription. Third, control is activated by access to resources. Partly, of course, this is a function of income, but it also relates to the facilitation of resources and the means of accessing them at the requisite level.

These three principles of limits, control and access might be useful as a means of measuring the actual extent to which choice is truly present in housing, always bearing in mind that these principles are to be seen as ideals and their attainment will never be absolute. Limitation and control are always going to be conditioned by questions of degree, and the control over resources will always, to an extent, be competitive, if only on grounds of scarcity. So they operate within the realm of owner occupation, where property rules limit the role of government, as well as that of other agencies and individuals. But what this notion of effective choice also suggests is that choice is not an intrinsic quality – it is not something we have within us – but is rather a condition determined by the constraints placed upon us and the level of resources at our disposal.

This leads to a final point to consider regarding choice, which is whether, how, and when choice is appropriate. As we have seen, when we raise the issue of resources, it becomes clear that choices will be limited. But we also need to factor in just who it is that is likely to be making the choices. Middle-class households with a steady income and savings in the bank are more capable of making choices and bearing the consequences of taking a bad option than a low-income household facing the prospect of homelessness. Similarly, the actual conditions in the local housing market are also important. If there is excess demand for private rented housing, and if social housing is scarce, then it might not really be appropriate to talk about choice. The only agents with choice in this situation are the landlords, in that they can decide who they want as tenants.

Choice will always lead to a situation of winners and losers. Some households will obtain exactly the housing they desire, whilst others will face a much more limited choice. Of course, virtually no one ever has an unlimited choice. We are always limited by our income, by the time we can afford to take and what is available. But the consequences of no choice or a poor choice might be quite considerable for some people, particularly those with low or no income of their own.

This raises the question of whether we are prepared to allow households to pay the consequences of poor decision-making. If we are allowed to choose, there is always the possibility that we will choose wrongly or badly. But is this acceptable in a welfare system aimed at protecting the vulnerable? If a household uses their housing allowance to buy alcohol and they are accordingly evicted, are we happy as a society to say that this is their fault and leave them to deal with the consequences? But, if we find that they used the rent money to pay a fuel bill, or buy a pair of school shoes for their child, would we take a different view?

This means that we have to be aware of the purpose of social housing and government subsidies. If housing subsidies exist to make housing more affordable and to allow individuals access to good quality housing, do we really want a system that allows some people to fail? Is the purpose of social provision to provide a safety net or is it a means of acculturation and education? Are we aiming to improve people morally by making them more responsible and capable, or are we more concerned to help them?

What this boils down to is the question of whether choice and welfare go together. This is, of course, a huge issue, and providing an answer is beyond the scope of this discussion. However, what it does show is that choice might be problematic and we need to be aware that certain consequences may accrue when society expects individual households to take decisions for themselves. This does not make choice a bad thing, but then nor should we automatically see it as a virtue.

Think piece: Nobby the street dweller

This is the story of a homeless man named Nobby, whose choice of residence was a bus shelter in a pleasant middle-class residential area. Nobby was a bad-tempered man who refused to say what his real name was or where he came from. He lived with all his possessions in the bus shelter and refused to leave. Initially he was resented by the local residents, but in time they came to accept him and even provide him with food, old clothes and so on. Indeed, when the police and the local authority tried to move him on, the local residents objected and stood up for Nobby's right to live where he chose. Indeed, he even refused an offer of rehousing by the local authority, choosing instead to maintain his current lifestyle. As far as he was concerned, he should be able to live as he liked and on his own terms: society could either ignore him or it had to bend to him.

- Should Nobby be able to choose to live as he wishes?
- Are the authorities being responsible by letting him?
- Should someone's choices ever be limited, even if no one else is being harmed?

Discussion points

1 If owner-occupiers can choose where they live, why shouldn't those who rent?
2 Should households be left to bear the consequences of their poor choices?
3 Is choice compatible with welfare?
4 Is choice really that important?

See also

Chapter 7 Need
Chapter 13 Desire
Chapter 16 Welfare
Chapter 22 Sources of finance
Chapter 23 Markets
Chapter 34 Planning

Further reading

Brown, T. and King, P. (2005): 'The Power to Choose: Effective Choice and Housing Policy', *European Journal of Housing Policy*, 5, 1, pp. 59–75.
Elster, J. (Ed.) (1986): *Rational Choice*, Oxford, Blackwell.
King, P. (2003): *A Social Philosophy of Housing*, Aldershot, Ashgate.
Mulder, C. (1996): 'Housing Choice', *Netherlands Journal of Housing and the Built Environment*, 11, 3, pp. 209–232.

9 Rights

One of the main justifications for government intervention is that people have rights. This means that there are certain things that they can expect and should have. Therefore, the importance of rights-based theories is that they offer foundational arguments for social action. But they can also be specific in that we attach a rights claim to a particular good or issue, such as housing or health care. Moreover, rights are used to argue both for state intervention and against it, such as the right to privacy or personal autonomy.

This latter point is important because theories of rights teach us about the importance of individual interests and that they cannot simply be overridden for a seemingly superior (social) good. As individuals, we want to believe that we are governed by our own thoughts and that we can have some significant control over our own lives. We are all important beings and we cannot – or should not – be reduced to mere servants of the interests of others and used as part of a trade-off between one set of interests and another. If we are all unique rights-bearing individuals, then in what sense can it be legitimate to sacrifice the interests of one or some of us for the perceived benefit of others? This might be seen as selfishness but, as soon as we realise that we all have rights, we can perceive it as something more substantial and worthy.

Therefore the significance of a rights-based argument is that it suggests that each individual person is important. Rights are what individuals possess, and they possess them by the simple fact of being human. Even so-called social rights, being the rights to certain socio-economic claims, are held by individuals. The importance of rights, therefore, is that they locate significance at the level of individuals and prohibit any trade-off between individuals and groups in which the interests of some are sacrificed for the benefit of others. What rights-based arguments help us to do is to concentrate on the morality of a situation rather than questions of utility or economy, and they allow one to argue that scarcity is not a sufficient condition for decision-making in housing policy.

But first we need to define briefly what rights are. Statements about rights are usually stated as a formula such as 'A has a right to do B'. What this statement means is that others have a duty not to prevent A from doing B. The point of such a duty is to promote or protect some interest of A's. Furthermore, even though this involves self-interest on the part of A, he or she should feel no embarrassment in insisting upon and enforcing this duty. Thus, a right is a legitimate claim that one person can make against others.

So, rights are concerned with how individuals are treated as differentiated persons. But this also suggests that rights are always expressed in regard to some other entity. One does not merely have rights, but rights *to* something. Moreover this 'something' would normally be deemed to be significant and of importance to the individual and to the society of which he or she is a part. Michael Freeden (1991) therefore defines a right as:

> a conceptual device, expressed in linguistic form, that assigns priority to certain human or social attributes regarded as essential to the adequate functioning of a human being; that is intended to serve as a protective capsule for those attributes; and that appeals for deliberate action to ensure protection.
>
> (p. 7)

We might suggest that rights depend on three conditions. First, that there are good reasons for demanding the thing, for example, legitimate ownership of a good, preservation of life and limb, and so on. Second, that there is something to enjoy (we do not have a right unless it relates to something, be it liberty, housing, health or whatever). Third, that certain social arrangements have to be made to assure or protect the possession of that something. This means that rights are never truly abstract in that they must relate to substantive entities and social relations.

While rights are enjoyed by individuals, it is also important to note that they are social relations. We only have rights because there are others that recognise and respect them. This raises the crucially important point that rights are correlative to duties. This means that to talk of rights is also a way of discussing people's responsibilities. Rights without a commensurate responsibility or duty are meaningless.

Rights are normally discussed in universal terms, and this means that they are reciprocal. Therefore A's right implies a duty on X, but X's right in turn imposes a duty on A. We all have rights and this means that we all have a side-constraint attached to our actions. This raises the important point that rights cannot be limitless, in that we have to moderate our interests in order not to infringe the interests of others. A person may pursue their own interests, but only to the extent that this does not infringe the rights of others. This implies that the pursuit of our interests involves both negotiation with other rights-bearing agents and some social mechanism to adjudicate when rights clash. It also suggests that the side-constraints on our actions might be considerable.

What follows from this brief attempt to define rights is that they imply a respect for interests from the point of view of the individual who has them. Our rights are not a matter of indifference to ourselves and should not be to others. As we are of intrinsic worth, we should be left to determine our own interests and not have them dictated to us. Our interests therefore are what we say they are and are based on the reasons that we give for them.

This should not be taken as meaning that whatever an individual describes as his or her interest must be respected. One must seek reasons from the person, and one would apply some test of significance (for instance, to adjudicate between the right to shelter and the right to fast cars). However, the main point remains, which is that interests are best self-described and should not be imposed.

It is common to categorise rights in one of two ways. First, is to see them as positive claims to actual goods and services. Rights can therefore be seen as socio-economic claims. So we have the right to health care, to education and, of course, to housing. We can come up with a list of these rights-claims which together add up to a civilised life or would constitute human flourishing. An important element in any discussion on socio-economic claims is just who is responsible for meeting them. As we have suggested, to say that I have a right means that you have a duty to fulfil it. But who is the appropriate person to fulfil my right for housing, health care and education? This involves expertise, resources and a competent assessment of what my needs are. I cannot therefore just go up to anyone in the street and ask them to house me. This suggests that, to seriously discuss socio-economic claims, we also need to consider what agencies are capable of fulfilling them, and in most societies the agency is either the state itself or one that has been empowered by the state. Indeed, it is often the case that the state is both the upholder of our positive rights as well as the agency charged with meeting them.

However, for the state to be able to fulfil its duty, it needs resources and this can usually only be obtained by taxing its citizens. Therefore the state uses compulsion – a tax is necessarily compulsory – in order to ensure that it has the resources to provide for the socio-economic claims of its citizens. But might this not imply a restriction on the rights of citizens? I presumably have a legitimate claim to my earned income and my property. It is mine to do with as I wish because I own it. Yet I now find that the government has confiscated a portion of it to meet the needs of others. This raises the second manner in which we can describe rights. Instead of seeing them as positive, where we come up with a list that the state should guarantee, we can see rights as negative, in that no one has the right to coerce me into actions not of my choosing. I should be left free to use my property and my abilities as I see fit rather than being directed by anyone else. This view of rights, then, does not describe what I should have in order to flourish or lead a civilised life, but rather it suggests that I should be left alone to use my talents and resources as I see fit. Hence it is common to refer to these as freedom rights.

It is not difficult to comprehend that these two views of rights are often seen as being in conflict. One the one hand, we have a list of what makes for a good life, which needs to be paid for by someone and, on the other hand, we have a view that says that individuals are best able to determine what is best for them. Thus, to take the property of someone means that they might not be able to fulfil their legitimate rights. So it might be argued that the more we seek to meet socio-economic claims, the more we erode freedom rights.

This is relevant to housing in that the right to property is one of the main freedom rights that we may experience. The property is ours because we can prove legitimate ownership. So what right has the state, or anyone else, to dictate how we use it? Is it proper for the state to regulate our use of it, to tell us what we can do with it and how much we can charge others for letting them use it? There is an apparent clash here in that the rights of a landlord to use her property as she desires might be in conflict with the socio-economic claims of a household who would otherwise be homeless.

An interesting way around the apparent clash between freedom rights and socio-economic claims has been offered by Jeremy Waldron (1993). In his essay, Waldron explores the nature of street homelessness and public and private property ownership. He shows that certain basic functions, such as sleeping, washing, urinating, etc., can and must be seen as freedom rights. He argues that we cannot undertake any sort of a life unless we can carry out these basic human functions. Yet these rights might not be exercisable in situations where property rules are rigidly enforced.

Property rules determine where one has a right to be. They define rights of use and exclusion. Thus, they grant the owner the power to exclude those with whom they do not wish to share the property. This situation applies whether the property is owned privately or by some public body. This means that some agents have property rights that they can legitimately exercise, and this may involve excluding all others from that property.

However, all actions are situated in that they must be done *somewhere*. One must sleep somewhere, wash somewhere, urinate somewhere and so on. Thus one is not free to perform an action unless there is somewhere where one is free to perform it. Waldron limits his discussion of actions to those absolutely necessary for human survival. However, his list is not an exhaustive one. Indeed, all actions, be they urinating, lovemaking, reading a book or discussing philosophy, are situated.

Homelessness is seen here as where one is excluded from all the places governed by private property rules. The homeless are entitled only to be in public places. They have no right to be on private property unless given permission by the owner. They must therefore rely on public places to undertake their situated functions. But this is possible only so long as the public authorities that own this property tolerate them. Just as private owners can exercise their right to exclude, so can public bodies. Waldron rightly points out that there is an increasing regulation and policing of public property that prevents the homeless from exercising their basic functions in public. Waldron gives the example of removing seating from subways in US cities to prevent them from being used by the homeless. This form of 'zero tolerance' of vagrancy can also be seen in the attitudes of politicians and public agencies in many other countries. The homeless are seen as having no right to be on the streets, and begging is seen as aggressive and intimidatory behaviour.

Waldron argues that a person not free to be in any place is not free to do anything. One important consequence of this argument is to show that freedom rights do indeed clash with each other. Property rights, as commonly defined in terms of exclusivity of use and disposal, are clearly freedom rights. Private individuals and public corporations who prevent the homeless from accessing their property are thus acting entirely legally and within their rights. Yet there are certain rights that we must have, homeless or not, if we are to carry out our basic functions. These too are freedom rights, in that we must be free to be in a place before we can undertake these basic functions. But the situated nature of this freedom means that certain rights can only be fulfilled when the property rights of some are overruled. Likewise, side-constraints prohibiting interference to property rights may well mean that the basic rights of others are infringed because they do not have the freedom to be. The homeless might be so constrained that they are literally unable to do anything without infringing the rights of others.

What is significant here is that Waldron is not casting the rights of the homeless as a socio-economic claim. They are not described as claims for housing which, being a finite resource, would involve competition between rival claims. Instead Waldron presents the case for the homeless in terms of a right to personal freedom. Therefore the 'right to be' is portrayed as having the same fundamental character as property rights.

Thus, to generalise from Waldron's argument, in order for us to undertake certain functions, there must be at least at one place where we have the right to be. This is based on the common-sense notion that human life is simply not possible unless these functions can be undertaken. These must be seen as rights, as legitimate claims on others, and therefore it follows that we need a place in order to undertake them. Thus the right to be in at least one place follows from the right to undertake certain basic functions.

What this discussion shows is that the distinction between positive and negative rights is not as hard and fast as some might claim. But, more importantly for our purposes, it points to the fundamental nature of housing. Describing housing as a freedom right does not diminish it as a socio-economic claim, but rather adds to its categorical importance. We might actually say that the connection of these two forms of rights shows that housing is quite fundamental and that it might actually serve as the basis for many other activities: if we haven't access to secure housing, just what else can we do?

Discussion points

1   Do we have a right to housing?
2   How do we decide on an outcome when rights conflict, for example, between a landlord and tenant?
3   Consider the view that rights are meaningless unless there are the resources there to fulfil them.

See also

Chapter 10 Responsibility
Chapter 12 Property rights
Chapter 19 Inequality
Chapter 20 Homelessness

Further reading

Freeden, M. (1991): *Rights*, Buckingham, Open University Press.
King, P. (2003): *A Social Philosophy of Housing*, Aldershot, Ashgate.
Waldron, J. (1993): 'Homelessness and the Issue of Freedom', in: *Liberal Rights: Collected Papers, 1981–1991*, Cambridge, Cambridge University Press, pp. 309–338.

10 Responsibility

If we make a decision, does that always mean that we have to bear the consequences? Clearly, if the outcome is a good one, then we will not hesitate to take responsibility. But what about if the decision goes wrong and we find we have made a mistake? Should we still be held responsible then? Many people might argue that the reasons for a situation are an important consideration in determining what is to be done to remedy that situation and who should be held responsible for doing it. If someone has caused a problem, then they should be the one charged with sorting it out.

However, what if that person, even though they accept their culpability, has no income or resources to sort out the mess? Take for example a household that has been evicted because of rent arrears. They knew that they had a legal responsibility to pay the rent and were also aware of the possible consequences. Therefore is it nobody's fault but theirs? But, what if this household includes young children? They were not party to the poor decisions of their parents, and so why should they suffer? So, even if we believe that the parents have behaved badly, we might still argue that what matters more is that they are homeless and they and their children need to be helped. What matters is not who caused the problem but how it can be sorted out and who is able to achieve this. A homeless family may be formally responsible for their situation but, once in it, they are not able to do much to help themselves. They lack the resources and the contacts to change their situation. We might argue, therefore, that what matters is that the family is rehoused and the children cared for, rather than blaming people for their past behaviour.

But there is a counter argument to this. If a household knows that they will be rehoused if they are evicted, why should they pay their rent? Might turning a blind eye to the reason for the eviction actually encourage irresponsible behaviour? Again, we might argue that this does not matter compared with helping a family in need. But, if resources are scarce, we will want to ensure that they are used as effectively as possible and this might include some calculation on the past behaviour of individuals as an indicator of how they might use these scarce resources in the future.

This discussion on responsibility centres on the notions of blame and task. So, how far should we blame people for their actions and use this as a means of adjudicating on future actions? Or should we be solely concerned with finding out who can sort out a problem and then tasking them accordingly?

Being responsible is where we are taken to be the primary cause of a particular situation. We are responsible because the situation would not have arisen but for our actions or omissions. This is to link responsibility with causality, in that we have to take upon ourselves certain tasks and actions as a result of past actions. This notion of responsibility is backward-looking, since it is how the situation has arisen that is considered important. As a result of this causality we are deemed to be the person (or agency) who is tasked with sorting the issue out: 'you caused it, so you sort it out'.

Robert Goodin (1998) suggests, however, that we should separate out cause and task. We can be held to be to blame for a particular situation that we have caused to come about, or we can be seen as the one tasked with its solution. It may well be that the fact that we are blamed will lead to us being tasked with its solution, but this need not be the case. What ought to matter, Goodin argues, is who is able to sort it out, rather than who caused it.

Goodin argues that blame responsibility is backward-looking and 'should be shunned for policy purposes' (1998, p. 150). This form of responsibility seeks to praise or blame people for what they have done in the past, even if the aim is to shape future behaviour. He favours task responsibility precisely because it is forward-looking, in that it specifies 'whose job it is to see to it that certain tasks are performed and that certain things are accomplished' (p. 150). Goodin admits that there will be a 'shadow of the past' even here, in that we might want to look at how a situation came about in order to allocate tasks. However, according to Goodin, this should be seen merely as 'a further consequence rather than the principal substance' (p. 151) of responsibility. What Goodin is saying is that, whilst a person might not be completely exonerated of blame, what matters is not who caused it but who can sort it out.

Goodin's reasoning for championing task responsibility is that history is not easily reversible and that 'people cannot always get themselves out of the jam as easily as they got themselves into it' (p. 152). He goes on:

> Sometimes others are better situated to get them out of a jam that they and they alone got themselves into. In such circumstances, who 'caused' the problem (whom to assign backward-looking 'blame responsibility' for it) is one thing. Who can best remedy it (whom to assign forward-looking 'task responsibility' for it) is another thing altogether.
>
> (p. 152)

Thus we may be homeless because we have omitted to make regular rent payments, and so we can be held to blame for this, but it does not mean that we are able to remedy our situation. Some other person or agency might be better placed to remedy this situation and thus be tasked with finding housing for us.

As Goodin himself recognises, this is a contentious argument, in that it might be supposed that, in some situations, the person responsible could be determined merely on the basis of relative resources, rather than on any moral basis at all. What is important to Goodin, however, is the empirical or instrumental quality of welfare. If we assert that a particular level of welfare or flourishing is desirable or even necessary, we should then be primarily concerned with achieving those outcomes: hence, the concern to look forward rather than back. Accordingly, he argues that smokers 'have only themselves to blame for their cancers' (p. 153). But once they have cancer, there is 'nothing further that they could do to cure themselves of it' (p. 153).

On one level, Goodin's point is correct, in that what is important in cancer treatment is not why the cancer is there and who is to blame, but how and whether it can be treated. We are more concerned with cure than cause. However, there is still a problem with Goodin's argument. He is prepared to admit that there is 'a shadow of the past' looming over the apportioning of tasks. Yet, apart from this, he suggests that blame and task are entirely separate. This may be the case, but that does not imply that we should adopt one view at the expense of the other. We may not be able to treat our own cancer, but we can still be blamed for it and held to account as well, and that is precisely because we know that smoking can cause cancer.

We can argue that certain forms of treatment should be seen as imperatives for which the state will pay, whilst others are to be met through private insurance. Thus we could exclude certain treatments from state-run health care systems and force individuals to insure themselves against them. Likewise, private insurance companies or even social insurance systems might choose to charge some customers more because of their behaviour. Smokers, for instance, pay higher life insurance premiums, and young men, who are more likely to be involved in road traffic accidents, pay more for car insurance. This does not mean that the insurance companies are shirking their 'task responsibility' if there is a cancer or a car accident. Rather it means that the risk is altered as a result of the actions of the customer, and it is felt to be acceptable that this risk be accounted for. In this sense the insurance company is quite legitimately incorporating the 'shadow of the past' into their deliberations. This is precisely because the shadow is deemed to be cast into the future and will materially affect the relation between the company and its clients.

We need also to be aware that *not* apportioning blame is a substantive issue in itself. By openly stating that we will not apportion blame, we might alter behaviour for the worse. As we have stated already, why should anyone act responsibly if they know that they will not be held to account for it? If this view is correct, it does matter who is held responsible and what is expected of them.

In the light of this concern, one way of discussing the notion of responsibility is whether it is internalised or externalised. David Schmidtz (1998) suggests that responsibility is externalised 'when people do not take responsibility: for messes they cause, for messes in which they find themselves' and when 'people regard the clean up as someone else's problem' (p. 8). This is precisely when we do not apportion blame, but instead state that responsibility should be located with an agency with the resources to solve it. The responsibility is therefore vested outside the individual who caused the problem.

On the other hand, Schmidtz states that responsibility 'is internalised when agents take responsibility: for their welfare, for their futures, for the consequences of their actions' (p. 8). Like Goodin, Schmidtz does not seek to relate responsibility to blame, seeing the issue as one of welfare or 'what makes people better off'. He argues that internalised responsibility is precisely what makes people better off, whilst externalising responsibility creates dependency and poverty: 'What strikes me about citizens of prosperous societies, then, is not their individualism so much as their willingness to take responsibility' (p. 9).

Schmidtz makes the point that, if we are not taken to be responsible, then this means that the responsibility lies with some other person or agency instead. Whilst it may be that we are not held completely at fault for causing the situation, it is still perverse to pass that responsibility on to society, 'that is, people who are not even partly at fault' (p. 11). A household evicted for not paying their rent might not be able to find new housing for themselves, but this does not of itself explain why taxpayers, who have nothing to do with this household, should now be held responsible for funding a replacement dwelling.

This leads Schmidtz to argue that 'what is woven into the welfare state is literally a pattern of transfer, not a pattern of sharing' (p. 75). Institutionalised welfare, he argues, benefits some people at the expense of others. Exonerating people from blame for their situation, in this sense, does not help make them responsible, but then passing the task on to those whose problem it is not does make *them* properly responsible. What happens is that individuals do not feel any sense of responsibility for their situation. But then, because the state undertakes to fulfil their welfare needs, individual taxpayers need feel no connection or sense of solidarity with those being assisted. Indeed they might even come to resent the state using 'their' money to support people with whom they feel that they have little in common.

Schmidtz suggests that responsibility becomes externalised because the role of institutions is seen as the alleviation of immediate problems rather than internalising responsibility. Schmidtz sees the problem as a static perspective in which only outcomes are considered. Instead, institutions should be concerned with the processes which individuals can use to fend for themselves. He suggests that 'property rights are pre-eminent among institutions that lead people to take responsibility for their welfare' (p. 22). He goes so far as to suggest that 'institutions of property (are) the human race's most pervasive and most successful experiment in internalised responsibility' (p. 25). What internalises responsibility more effectively than anything else is the attribution of property rights to individuals.

This is a controversial argument, but it is one that has recently had some resonance with policymakers. In particular, we have seen attempts in some countries to introduce a degree of conditionality, so that benefit recipients only receive money if they agree to undertake training or community work that prepares them for employment. The idea behind these reforms is that structures of provision should be geared towards changing the behaviour of individuals rather than merely keeping them on benefit. Benefit recipients should be forced to be active in their search for work rather than passively accepting welfare. This, as Schmidtz suggests, is a structural problem in terms of how welfare systems are organised.

What this discussion does show, however, is that we cannot hope to achieve a situation where some part of what we are doing is not contested. Both the arguments of Goodin and Schmidtz have merit to them, even if one may appeal more than the other. We know that people respond to incentives and will alter their behaviour if they think it will benefit them. But we also know of the severe effects of making a poor decision and that there ought to be an easy way out, particularly for households on low incomes. This means that there will be argument over the proper role of the state and what we should expect of individuals. It also suggests that there may not be a complete answer available to us, and so the arguments will continue.

Think piece: The limits of a landlord's responsibility

Mr and Mrs Smith are divorced as a result of Mr Smith's unreasonable behaviour. They were social tenants living in a two-bedroom house along with their teenage son. Mrs Smith now has custody of the child, but Mr Smith has access rights which include his son staying with him.

Mrs Smith has been allowed to maintain the social tenancy, but Mr Smith has not yet moved out. This is because he is demanding that he be allocated a two-bedroom property so that his son can stay with him when he has access. The landlord has only offered a one-bedroom flat on the grounds that this is consistent with their allocation policy because Mr Smith will be on his own for the majority of the time. Both Mr and Mrs Smith consider their landlord to be acting negligently and unsympathetically to their plight. Accordingly, Mr Smith stays in the house.

- Is the landlord acting properly here?
- Who is responsible for sorting out this situation?
- Does the fact that Mr Smith caused the separation make any difference to who sorts out the problem?

Discussion points

1 Does it matter how someone got into a mess?
2 Should help to the homeless be unconditional?
3 How much responsibility do we have for each other?

See also

Chapter 9 Rights
Chapter 12 Property rights
Chapter 20 Homelessness
Chapter 30 Government

Further reading

Goodin, R. (1998): 'Social Welfare as a Collective Social Responsibility', in Schmidtz, D. and Goodin, R: *Social Welfare and Individual Responsibility*, Cambridge, Cambridge University Press, pp. 97–195.
King, P. (2003): *A Social Philosophy of Housing*, Aldershot, Ashgate.
Schmidtz, D. (1998): 'Taking Responsibility', in Schmidtz, D. and Goodin, R: *Social Welfare and Individual Responsibility*, Cambridge, Cambridge University Press, pp. 1–96.

Part 3

Tenure

11 Owner occupation

Owner occupation is often portrayed as the tenure of choice. It is what households aspire to and what politicians focus on. Of course, in many countries there is a large and vibrant private rented market, offering a range of accommodation of widely differing price and quality. But what tends to create both economic and political waves is the fact that, in many European countries and throughout the English-speaking world, owner occupation is now the dominant tenure and the tenure that most households expect and aspire to.

The significance of the dominance of owner occupation became all too clear with the 2008 financial crisis caused, in part, by problems in the sub-prime mortgage market in the US and the securitisation of mortgage debt, which led to the subsequent world-wide credit crunch. In response to the financial crisis, governments across the world were forced to act to shore up banks and mortgage lenders, and to deal with the effects of the recession that followed the credit crunch. Many governments have been forced to cut public spending, and in many places housing markets have struggled to recover. Yet, despite this, there appears to be no attempt to suggest that owner occupation is not a legitimate aspiration, and consequently governments have continued to promote and support the tenure. It appears that the only alternative to a housing market that does not work well is one that does.

The issue for governments is that the scale of owner occupation is such that they cannot afford to leave it alone. With rates of owner occupation between 60 and 70 per cent in many countries, what happens to housing markets is a matter of concern to government. But also, because of the size of the tenure and the fact that housing is typically the largest item in a household's budget, governments are aware that they can affect consumer behaviour through the manipulation of housing costs. For example, a government can use its ability to control interest rates to affect mortgage costs with the aim of reducing consumer spending and thereby reducing inflationary pressures on the economy.

But it might be argued that the reason that owner occupation is so popular is because it is supported by government. The US government, for example, has been providing mortgage guarantees since the 1930s, while we can argue that the most significant housing policy in the UK over the last 30 years has been the Right to Buy, allowing social tenants to purchase their dwelling at a discount. Whilst politicians tend to do things that they think will be popular, we can also state that things tend to increase in popularity if they are subsidised by government. Unravelling this situation – is owner occupation popular because it has been subsidised, or subsidised because it is popular? – is virtually impossible. What we can be sure of is that owner occupation is too important to ignore.

There has been a long and inconclusive debate in academic circles about the connection between tenure and voting behaviour. In the UK, following the introduction of the Right to Buy in 1981, there was considerable debate as to whether promoting owner occupation encouraged people to vote Conservative. While this debate is perhaps now moot – the Conservatives did not manage to gain a majority in the UK between 1992 and 2015 – the politicians themselves certainly believed that such a link existed, and this perhaps helps to explain why there has been no serious attempt to abolish the Right to Buy.

But the support for owner occupation goes beyond political popularity. There are a number of economic reasons for the support that government offers to the tenure. First, as we have suggested, housing costs are usually the largest single item in any household's budget. This means that changes in housing costs are likely to have considerable effects on total household consumption. A general increase in housing costs, through an increase in the costs of borrowing, has major economic effects through reducing general consumption. Importantly, these effects go beyond housing and affect demand in other markets, as households have less disposable income after housing costs. An increase in mortgage costs reduces the income that a household has to spend on other activities, like holidays, motoring, and new electronic and white goods, etc.

A further effect of owner occupation is the so-called cascade effect, whereby housing wealth is passed from the one generation to the next. Most children will inherit the property of their parents and this effectively liquidates the wealth stored in the property. This means that the wealth of a significant number of households, most of whom are already owner-occupiers, is further enhanced. This may well mean that they consume more housing, and the demand for larger dwellings increases. Thus, the wealth generated by owner occupation tends to become polarised, with some households becoming quite asset rich as a result of inheritance, whilst renters (who are also often following a generational pattern) miss out.

Third, we can suggest that owner occupation is important economically because it results in a considerable amount of derived demand. This is market activity that arises as a result of transactions in the housing market. Many markets depend upon an active housing market. For example, the demand for DIY products, home furnishings, white goods and insurance all depend on a vibrant housing market. In addition, many solicitors depend on their conveyancing business, and estate agents are particularly susceptible to changes in the market as they rely entirely on the level of sales. A boom or slump in the housing market has considerable spill-over effects and thus owner occupation should not be taken in isolation.

Fourth, housebuilding has traditionally been seen as a means of boosting economic activity. This is because, as in all areas of construction, housebuilding is labour intensive and not particularly import sensitive. Thus, an active housing market, which encourages new development, can have a positive effect on local labour markets.

Fifth, housing is a store of wealth, in that its value tends to increase over time. However, a household's housing costs, assuming interest rates remain stable, does not change overmuch. Over time, therefore, there may well be a significant divergence between a household's regular costs (i.e. mortgage repayments) and the value of the dwelling. This can create a considerable benefit to households in that they may be able to tap into the free equity of their property to fund additional expenditure. This is known as *equity withdrawal*. This will have the effect of increasing consumption and might help to boost economic activity. However, in other circumstances it might be inflationary.

But this divergence between cost and value can also work in the opposite way. In periods of falling house prices there may be some households who find that the value of their property is now less than their mortgage. This negative equity will often mean that the households cannot sell their dwelling without making a loss. This problem is often compounded by the fact that, as housing costs rise, perhaps due to increasing interest rates, house values decline because owner occupation becomes less attractive. Indeed, the level of interest rates has an important effect on the cost of housing, which operates independently of the initial cost of purchasing the dwelling.

One of the most important issues is when the property was purchased. It is perfectly possible for one household to be paying twice as much as their neighbour for a similar property. This is simply because of when they bought it. If a household took out a mortgage of £50,000 to purchase the average house in 1990, then they are still making repayments based on that cost, even though the value of that average house in 2015 might be £200,000. But anyone purchasing the average dwelling in 2015 would be expected to repay a mortgage based on a property valuation of £200,000 plus.

The fact that housing costs change over time highlights another important facet of owner occupation. Not only is a property the most expensive item we are ever likely to buy, it is also the only commodity we will purchase without knowledge of its final cost. This is because factors such as interest rates and government policy might change over the life of a mortgage. Over the life of a 25-year mortgage, interest rates will have gone up and down and government policies may have changed, perhaps with changes to subsidies that directly affect the cost of housing.

This brings us to the issue of government support for owner occupation. Often the rhetoric behind the tenure is of choice, independence and personal responsibility. Yet this hides the fact that owner occupation has received considerable state support. As we have seen, the US government has helped to underwrite the mortgage market since the 1930s, and governments in Europe have also sought to regulate housing finance.

The most common form of support provided by government is either in the form of mortgage guarantees to lenders, aimed at offering some security to lenders, or through tax relief allowing households to offset part of their mortgage payments against tax. These may be targeted on particular groups as with the Low-Income Housing Tax Credit in the US, but tax relief might also be rather less discriminating, as was the case with Mortgage Interest Tax Relief, which was offered in the UK until 2000. This form of tax relief was not made on the basis of housing need or low income, but was, rather, universal, in that all households paying tax and with a mortgage were entitled to it.

This raises a very important aspect of government policy towards owner occupation. Unlike other support that is justified on the grounds of dealing with severe housing need or to support the vulnerable, these subsidies are often paid out regardless of income. They exist to make owner occupation more appealing for the general population. This returns us to the political support for owner occupation and its place as the dominant tenure. We can suggest that owner occupation is too important for politicians and policymakers to ignore and that, as a result, it will most likely remain at the centre of housing policy.

Think piece: US government supports owner occupation

The popular image about owner occupation is that it encourages independence and personal responsibility. But governments, for whatever motives, have been keen to support the tenure, and we might say that in some countries it has received greater support than social housing. Such is the situation with the United States. Niall Ferguson shows that the promotion of owner occupation began in the post-Depression period. He characterises housing finance before the 1930s as being based on short-term (3–5 years) interest-only mortgages, which meant that the mortgagee faced a very large final payment or would need to remortgage at regular intervals. The depression after 1929 had a disastrous impact on this market, with a precipitous fall in land values and massive increases in unemployment. As a result many households lost their dwellings.

In response to this the Federal government chose to intervene to create a more stable mortgage market. This was achieved by allowing for the development of long-term mortgage finance and by underpinning the market through a system of deposit insurance. The Roosevelt administration created the Federal Housing Administration, which provided federal insurance for mortgage lenders and encouraged long-term (20-year), fully amortised, low-interest loans. In effect this provided a form of nationwide standardisation and regulation of the mortgage market in the US. Further impetus was given by the creation of the Federal National Mortgage Association – known as Fannie Mae – with the role of issuing bonds and using the proceeds to buy mortgages from local lenders. In return, the Savings and Loans were restricted to providing finance to depositors at low rates. The result of this institutional backing was a mortgage market based on fixed long-term interest rates and security of deposits. Accordingly, it is no exaggeration to state, as Ferguson does, that the 'the US government was effectively underwriting the mortgage market, encouraging lenders and borrowers to get together' (2008, p. 249).

Reference

Ferguson, N. (2008): *The Ascent of Money: A Financial History of the World*, London, Allen Lane.

Discussion points

1   Are subsidies to owner-occupiers ever justified?
2   Is it acceptable to use other tenures to support owner occupation?
3   Is owning natural or does it depend on particular cultural conditions?

See also

Chapter 5 Ideology
Chapter 22 Sources of finance
Chapter 25 Housing allowances
Chapter 27 Boom and bust
Chapter 28 Borrowing

Further reading

Saunders, P. (1990): *A Nation of Home Owners*, London, Unwin Hyman.

12 Property rights

When we say that we own something, what does this mean? What does it mean to say that I own my own house? And how does this differ from renting? When someone rents a property, they are given exclusive use over it. This may be for a limited period of time or it may be open-ended, but the exclusivity is largely the same as if we own the property. We have a legally binding agreement with the landlord that we can have sole use. Renting, therefore, is where we have been given certain rights over the property. The same applies when we purchase a property: we too have rights over it. So just what are property rights?

Essentially property rights involve three things. First, to say we own something means that we have the right to use it. It is legitimately at our disposal and we can determine what is done with it. This use is of course limited to the extent that it does not impinge on the rights of others or infringe accepted forms of behaviour: I can own a knife but that does not extend to the right to put it into somebody else's chest. The second element follows on from this idea of use in that it is assumed that the right is to exclusive use. It is for me to decide who else, if anyone, may use it, and I can quite properly prevent all others from having access to it. My rights are therefore backed by some form of protection which allows for the exclusion of unwanted others.

The third element, and this is what separates owning a house from renting one, is the right to dispose or transfer the item as I see fit. As it is mine, I may dispose of it as I wish. I can sell the property for as much as I can get for it, or I can give it away. I can transfer part or all of the property to another (perhaps to my wife or children for tax reasons) and I can leave it to whom I wish. Again there are legal caveats preventing me from acting in a way that harms others. So I may sell my house when I please, but I cannot refuse to sell it to a black person or a Jew, and nor can I dispose of it by burning it down.

A renter has rights of exclusive use, but they may not dispose of or transfer the property. This is simply because it is not theirs. A renter is given exclusive use in consideration of a user charge – rent – and within the bounds of the legal conventions of the country in which they reside.

But, to say that we own something means that we have the exclusive use of it and the right to transfer it to another. So, when we carry out a transaction in a market, we transfer our rights of exclusive use and disposal over something (which may be money) in return for gaining rights over something else. Property rights are therefore a necessary element of any market-based system.

However, property ownership is only meaningful if others recognise the fact. We have to be able to prove that we own something, and this has to be generally accepted within the society of which we are part. In most developed countries the owner can prove ownership of their house and car. They will have the deeds and registration documents. Moreover, there is a formal legal system that will back up their claims. If someone tries to take their car, they have a form of redress that either prevents that person from taking it or allows the owner to claim compensation and demand punishment. We have a visible sign of our ownership and this is supported by what the Peruvian economist, Hernando De Soto, calls a 'representational process' (2000, p. 7).

However, as De Soto states, 'Third World and former communist nations do not have this representational process' (2000, p. 7). This means that they cannot use their capital, their resources, as they might wish. As De Soto states:

> The single most important source of funds for new businesses in the United States is a mortgage on the entrepreneur's house. These assets can also provide a link to the owner's credit history, an accountable address for the creation of reliable and universal public utilities, and a foundation for the creation of securities.
>
> (p. 7)

In the USA having a system that recognises property rights allows for the creation of businesses and therefore more wealth. There is a clear framework of law for markets to operate.

However, in the developing world this representational process does not exist. In sub-Saharan Africa that ready system of accountability is not there, with the result that proving ownership of land involves a lengthy and time-consuming process, perhaps involving bribing public officials in order to obtain permission. De Soto shows that this is also a problem in Latin America, parts of post-communist Europe and other parts of Africa. The problem is that individuals and businesses have no straightforward and dependable means of achieving recognition of their rights to property.

De Soto argues, therefore, that the problem for developing countries is not a lack of capital but its recognition. He states that, even in poor countries like Haiti and Egypt, the poor save: 'In Egypt, for instance, the wealth that the poor have accumulated is worth fifty-five times as much as the sum of all direct foreign investment ever recorded there, including the Suez Canal and the Aswan Dam' (2000, p. 6). The poor, however, hold their resources in what he calls 'defective forms', for example, 'houses built on land whose ownership rights are not adequately recorded' (p. 6).

So it is not the lack of possessions that is the problem, but rather how they can, or in fact cannot, be used:

> Because the rights to these possessions are not adequately documented, these assets cannot readily be turned into capital, cannot be traded outside of narrow local circles where people know and trust each other, cannot be used as collateral for a loan and cannot be used as a share against an investment.
>
> (p. 6)

But all these things are possible in the developed world. Accordingly, De Soto by default demonstrates what a formal structure of property ownership achieves in a market: it provides a title to things – we can prove that they are ours; we can trust others and they can trust us because of the system that underpins our relations; and so we are able to use our property to improve our own situation and society as a whole.

A somewhat different view of property right is taken by conservative thinkers who see property ownership as forming the basis of the type of society that they wish to see. However, this is not just because of the role of markets, but also the fact that property rights create social relations: owning things, and the consequent need to transfer and exchange things, means that we have to relate to others around us. We might suggest, therefore, that operating within markets is the way in which societies are created and develop. This is not, however, because they maximise individual freedom but rather because they create social obligations and so strengthen the bonds within society. Property ownership, for conservatives, has the virtue of giving individuals a stake in society and therefore an interest in its maintenance.

This view of property can be seen in the thought of Roger Scruton (2001), who discusses what he calls our 'absolute and ineradicable need for private property' (p. 92). He justifies this ineradicable need by stating that:

> Ownership is the primary relation through which man and nature come together. It is therefore the first stage in the socialising of objects, and the condition of all higher institutions. It is not necessarily a product of greed or exploitation, but it is necessarily a part of the process whereby people free themselves from the power of things, transforming resistant nature into compliant image. Through property man imbues his world with will, and begins therein to discover himself as a social being.
>
> (p. 92)

As Scruton states, 'Through property an object ceases to be a mere inanimate thing, and becomes instead the focus of rights and obligations' (p. 93). Through property ownership 'the object is lifted out of mere "thinghood" and rendered up to humanity' (p. 93). It bears the imprint of social relations and reflects back to the owner 'a picture of himself as a social being' (p. 93), as someone now with the capability of relations with others. Property ownership is therefore seen by Scruton as a primary social relation. It is what allows us access to the social world, as beings able to achieve our ends.

This is a rather philosophical way of saying that, without property we cannot identify any object in the world as our own, and hence we have no right to use any object, nor can we expect others to allow us access to it (not that they could, of course, because they too would have no rights over it). Without rights of ownership, everything is merely an object of desire. Objects without ownership can play no part in social relations: there can be no exchange, no gifts and no transfers from one person to another.

Scruton argues that, if people are to become fully aware of themselves as agents who are capable of independent action within a social whole, then they need to see the world in terms of rights, responsibility and freedom. He suggests that it is 'The institution of property [that] allows them to do this' (p. 93). By making an object mine, I can now use it for my own purposes. I am able to be more active because my possibilities have been increased. But I have also been given a responsibility, for I now have to determine how it can be used, whether I should share my access, and so on.

Scruton therefore emphasises the social nature of property and markets by showing how it connects humans together and gives them responsibilities and the need to respect others: what is theirs is not mine, and what is mine is not theirs. If I want what is not mine then I must engage with others through negotiation, trade or persuasion, rather than brute force.

Having given this rather abstract justification of property ownership, Scruton then identifies the main form of property that we experience. Ownership, as it were, grounds the self into the social world. As he states, instead of being at loose in the world, an individual is 'at home' (p. 93). He goes on:

> It is for this reason that a person's principal proprietary attitude is towards his immediate surroundings – house, room, furniture – towards those things with which he is, so to speak, mingled. It is the home, therefore, that is the principal sphere of property, and the principal locus of the gift.
>
> (p. 93)

The most important form of property is the home, as this is the primary relationship with things in the world. It is what we live within and therefore becomes part of us. When we own those things around us – the house and its contents – we are better able to control our surroundings and fulfil our personal and social obligations. The family unit is where we show responsibility to others, where our primary obligations are held and where we are most able to express our generosity.

Scruton's argument can be clearly linked to the political support offered by conservative (and other) politicians for owner occupation. They see a distinct virtue in owning that goes beyond economic arguments but which rests on the ideas of social stability and personal responsibility. Owning property, they argue, changes the manner in which individuals behave, which, in turn, changes a society. This is a view that we can contest. Indeed it is hard to see owner occupation having a stabilising effect in many societies after the 2008 financial crisis. However, we need to recognise that this conservative view has had a particular resonance even if it is seldom expressed in the manner undertaken by Scruton.

Indeed, whether we subscribe to Scruton's view or not, we need to appreciate the importance of property rights to housing markets and to owner occupation as the dominant tenure in many countries. Without property rights, housing systems would be unrecognisable from what we now know and expect.

Discussion points

1   Consider the ways in which property rights differ between renting and owning.
2   What would a society look like without property rights?
3   Does owning things make us behave differently?

See also

Chapter 1 Housing and home
Chapter 5 Ideology
Chapter 11 Owner occupation
Chapter 12 Property rights
Chapter 23 Markets
Chapter 24 Rent
Chapter 27 Boom and bust

Further reading

De Soto, H. (2000): *The Mystery of Capital: Why Capitalism Triumphs in the West and Fails Everywhere Else*, London, Black Swan.
Scruton, R. (2001): *The Meaning of Conservatism*, 3rd edn, Basingstoke, Palgrave.

13 Desire

Households choose to become owner-occupiers. They do so because they desire it and because they feel they can achieve it. We might say that government is as much responding to this desire as leading or creating it. Indeed, this desire has been important in feeding the growth of owner occupation. Over the last two decades households in countries such as Ireland, Spain, the UK and the USA have become obsessed by property to the extent that it is glorified and objectified in an almost pornographic way. The TV channels seem to be full of programmes on purchasing properties, or on doing them up. Many of us sit transfixed as others go through the apparent agonies of deciding on the right property, whether or not they can afford it and then how to redesign it. Such programmes have made the careers of so-called 'experts', who are able to advise on design or lead naïve couples through the complexities of house buying. These programmes, with their telegenic presenters, offered the prospect that households could surf the housing boom, spot a bargain and improve it in a manner that would enhance its value (with the implication that it could be sold on at a profit).

This fetishisation of property is considered by Marjorie Garber (2000) in a book titled *Sex and Real Estate*. She sees that concern for houses and their interiors has replaced sex as the main preoccupation of many households. This is because she sees the same sense of desire for property and furnishing as for the bodies of those we love and desire. The subtitle of the book is *Why We Love Houses* and she indulges in some loving descriptions of kitchen worktops and soft furnishing. Garber suggests that we have affairs with our dwellings, falling in love with them and carrying on the affair with great intensity, before having our eyes turned by something else that comes along. She is somewhat ambivalent about this herself: on the one hand, she is an academic and able to look at the phenomenon with a degree of detachment; on the other hand, she too admits to being sucked into this circuit of desire, so that much of the material for her book is from her own experience.

This attitude of Garber's is rather typical of the ambivalence we see with regard to property, where it has an immense symbolic significance, whilst at the same time being a place that fulfils an existential need. Nowhere is this ambivalence more evident than in the generic name given to domestic design magazines in the USA. As Garber reports, they are referred to as shelter magazines, a designation which carries with it the notion of an imperative condition that could not be further from the concerns of these magazines, with their emphasis on contemporary design, and articles on luxury dwellings owned by the rich and famous. Indeed, for many people in the UK the term 'shelter' is now so closely associated with the most high-profile homeless charity as to make the use of the term in the USA seem almost offensive: what could be more inappropriate than the peddling of luxury on an existential condition?

One aspect, then, of this glorification of property is that we lose sight of what it is for. We might forget that it is a place to live in, to raise a family, to be intimate and close to those we love. The dwelling becomes more than a place, taking the form of an object of desire. What we have, when we link this to the support for owner occupation given by government, is a potentially toxic mix in which property is something we obsessively desire, but which is then promoted by government and subsidised precisely because it is so desired. Government sees it as an imperative to help us make the down payment on our dream; it is, after all, one we all share, and why should anyone be left out? But this support only legitimises the process, and this makes it easier for us to choose. But, perhaps, in doing so, we start to forget the consequences that often go with choosing.

Discussion points

1   Is housing just a commodity?
2   What do you use yours for?

See also

Chapter 1 Housing and home
Chapter 11 Owner occupation
Chapter 12 Property rights
Chapter 27 Boom and bust
Chapter 28 Borrowing
Chapter 34 Planning

Further reading

Garber, M. (2000): *Sex and Real Estate: Why We Love Houses*, New York, Anchor Books.

14 Social housing

There are two main ways in which the state can support those in severe housing need. It can provide them with additional income so that they can buy housing in a market (see Chapter 25, Housing allowances) or it can provide dwellings at heavily subsidised rents. The latter form of provision is what we refer to generally as social housing. The term 'social' clearly has a particular connotation, with the implication that it has been provided by society or has a social purpose. In this sense it is different from private renting, which is based on commercial incentives and the profit motive. Social housing is deemed necessary out of a sense of solidarity across a community or because society believes it has a responsibility for those who are vulnerable and unable to provide for themselves.

Social housing therefore exists for two overlapping purposes. First, it is to help those on low incomes and second, it is to ensure that this housing is built, managed and maintained to a particular level. So, the provision of social housing is about quality as much as it is a concern for quantity.

Social housing exists in different forms in different countries. In some places the state has itself built dwellings, as in the case of the UK and Ireland. In other places it is a matter of the state providing subsidies or loan guarantees to allow private providers to build housing of a particular state-determined standard. So, to refer to housing as 'social' does not imply a particular model or legal framework.

Indeed, we can define social housing in three ways. First, we can define social housing in terms of what it is for. Typically, this would be where we see need or vulnerability as the main justification for the provision of housing by government, as well as for its allocation. If we concern ourselves with the purpose of social housing, what matters therefore is who gets into the dwellings.

Second, we can define social housing in terms of who owns it. This particularly pertains to countries, like the UK and Ireland, which built up large stocks of social housing over much of the twentieth century. In the UK the issue of ownership has been very important in defining social housing, with the idea of local political accountability being held very strongly, at least until the late 1990s. We can see this in the debate over the Right to Buy, which involved the sale at a discount of over 2.5 million dwellings (more than a third of the total stock) to sitting tenants. This was seen by some as a form of privatisation and a denuding of national assets for private gain.

More recently the situation has become more complicated and not just because of the effects of the Right to Buy. Since the early 1990s, nearly a million dwellings have been transferred to either new or existing housing associations, and this has sat alongside a policy of all new social housing being built by housing associations. Therefore social housing in the UK is no longer council housing, and the issue of being owned by the state is now less relevant.

The third means of defining social housing is through who pays for it. Housing can be defined as social if it is funded by government with the aim of achieving some particular purpose. So we might take the use of government subsidies as a defining characteristic. However, social housing in several countries is funded by a mix of public and private finance, and, of course, tenants still have to pay rent, which may come from their private income.

Perhaps, then, the most useful definition relates to whom social housing is for rather than who owns it or how it is funded. Social housing aims to provide good quality housing to those who cannot provide it for themselves through a market. We can suggest that this means that social housing exists as a form of solidarity. Related to this, we can perhaps argue that social housing seeks to remedy social divisions by putting all households on a more equal footing. Therefore government intervenes in markets to ensure that all people are well housed, and historically has chosen the direct provision of housing in order to achieve this aim.

One of the main ways in which government supports social housing has been through the use of object subsidies. These are subsidies paid to landlords to enable them to build new housing and offer it at relatively cheaper rents. Object subsidies – sometimes also referred to as supply-side, capital, or bricks and mortar subsidies – are particularly effective in dealing with housing shortages. However, in many countries over the last 30 years there has been a tendency to reduce the use of object subsidies and replace these with subject subsidies paid to households. The reasons for this are partly ideological, with the political success of free market ideas in many countries, but it is also because of the perceived unpopularity of social housing compared with other tenures, particularly owner occupation. We might add to this the expense of mass housebuilding at a time when governments have become increasingly conscious of the need to restrain public spending and keep taxation relatively low.

However, there are a number of perfectly sound reasons for the direct provision of housing through the provision of subsidies to landlords. The first justification is that housing is a merit good and it is therefore socially desirable to provide good quality housing. Merit goods can be defined as those goods that society believes we should have but which we decide not to purchase. The consumption at a certain level might be good for us as individuals and for society as whole. It will lead to healthy individuals who are able to participate fully as active citizens. However, it might be the case that we do not perceive these benefits and so there is need for government to encourage citizens to consume housing at the socially optimum level.

Merit goods are therefore goods that individuals ought to consume at a certain level, because it is good for them. However, they may not be fully aware of their benefits or may choose not to consume to the desired level. Thus there might be a discrepancy between what individuals wish to do and what society as a whole thinks is best. Therefore, according to this argument, there might still be a problem even if individuals received enough money themselves with which to purchase or rent houses. They might not use the money to purchase sufficiently 'good quality' housing because they do not see the merit in doing so. The issue of merit goods, therefore, is as much about understanding and knowledge as it is money.

An argument in favour of social provision that is linked to the merit good argument is that paying individuals in cash might lead them to spend the money on other things, whereas funding the provision directly means that it is spent on the intended goods. Housing consumption is politically acceptable, whereas a cash payment, which could be used for such things as alcohol and tobacco, might not be. As a society we approve of certain activities as being legitimate for subsidy, but not others. Thus we might want to ensure that public money is spent on things which benefit individuals and not merely on wants and desires.

But, were we to accept this argument, we would offer no cash benefits whatsoever and merely provide clothing vouchers, food parcels and so on, to those who were not well off. Clearly society feels comfortable providing benefits and pensions for its citizens and feels sanguine about their competence to spend the money wisely. We might therefore argue that the same ought to apply to housing if individuals are capable of making decisions for themselves about how they spend their income.

We can point to a linked argument, which states that it is not fair to allow people on low incomes to make choices which can affect them disproportionately when compared with those on reasonable incomes. Thus, to give individuals a housing allowance and tell them to pay their rent, as well as food, transport and fuel bills, is to set them up to fail. Therefore, the argument runs, it is better to provide houses rather than housing allowances.

Another argument for social provision is that poor quality housing can lead to wider societal problems such as ill health, vandalism, racism, family break-up, etc. If people live in poor quality housing they may become ill or, in the event of a shortage of suitable housing in an area, there might be racial tensions if some groups believe that they are being excluded and others given preferential treatment. The point is that housing can have far-reaching effects, which go beyond fulfilling the wants of individual households. Housing provision, or the lack of it, can have wider social effects, and it is difficult for individuals to deal with these problems themselves. Therefore it is suggested that building more social housing, and to a high standard, can help to deal with these social problems.

Yet this argument only works if there are no resulting social problems in social housing. Unfortunately this is not always the case, and many social housing estates have proven to be centres of economic dependency, anti-social behaviour and poverty. Social housing might be unpopular and attract a certain degree of stigma that mitigates the wider social effects of the provision.

A more technical argument for social housing is that, because of the differences in land and property values across the country, there are differential costs in rental markets, and these can be ironed out by the provision of social housing at controlled rent levels. In this way a government can ensure that there is a comprehensive coverage of affordable housing across the entire country. By subsidising the production of social housing, even if it means paying higher levels of subsidy to landlords in high-cost areas, rents can be similar across the country. This, it is argued, can encourage labour mobility, as well as being seen to be fair and just.

A problem associated with subject subsidies is that they can create a poverty trap because individuals are reluctant to take low-paid work because of the way in which their benefits are withdrawn as their income rises. It can therefore be argued that providing the goods 'in kind', in the form of social housing, would help to deal with this problem.

But this assumes that rents are low enough to ensure access to those on very low incomes. In practice, in many countries some social housing tenants can only afford their housing because they receive a subject subsidy. We might also question what impact zero or very low rents would have on incentives: why give up a good that is free, even if our income rises considerably?

One great advantage of object subsidies, for government if no one else, is that they can allow greater control over the quality of housing provided for low-income households. By providing subsidies to a particular level and applying a particular control and monitoring regime, government can ensure a high quality of outcomes. Conversely, it can also ensure that recipients do not benefit excessively from public funds. A key problem with housing allowance systems is that government finds it hard to control the number of recipients and rent levels, and therefore the level of expenditure. Government can limit entitlements and eligibility, but this is less fine-grained than the control that it can have over object subsidies, where it can much more accurately set the limits of funding.

The final, and perhaps the strongest, argument for object subsidies is that they act as direct incentives to supply new housing. It was argued that, if we have a shortage of housing, as was the case in most Western countries throughout much of the twentieth century, subsidising landlords is the most direct and effective way of getting houses built. Furthermore it also encourages quality by allowing landlords to build to a higher standard than they otherwise would. If left to a market, they would perhaps be more concerned with covering their costs and making a profit based on the limitations of the budgets of the households to whom the houses would be let. Of course, this presupposes that government is capable of planning effectively so that housing is in the right place and of the right price and quality.

These are strong arguments in favour of social provision and it has many advocates in politics, academia and the housing professions. Yet, despite this, social housing is in decline in many countries and is seen as a poor relation in terms of owner occupation. It is also the case that social housing tends to suffer, relative to other forms of social provision such as health and education. Funding for social housing is often an easy target for spending reductions. The reason for this is that, almost without exception, social housing houses only a minority of households in any country. It is not the dominant tenure. Moreover, access to social housing is often determined by a means test that tends to create a division between those who receive assistance and those who pay for it, who almost by definition would be considered too affluent to qualify for a social tenancy. It is therefore not surprising that social housing fails to be as popular as public services that are offered on the basis of universal provision.

It would be simple to argue that social housing is an anachronism or a throwback to a different age. It no longer has a place in an era where households expect to have a choice and enjoy the freedom to spend their income as they please. Now it may indeed be the case that social housing will never be the majority tenure. However, there will always be a number of households with no or very little income or who are not capable of looking after themselves for whatever reason. This suggests that there will always be a role for social provision of some sort. The question is over how extensive that role should be: is social housing merely a safety net or does it have a wider purpose?

Think piece: The supportive shack and the oppressive house

John Turner, a British architect working in Latin and Central America in the 1960s, demonstrates how government subsidies to provide social housing can actually prove to be unproductive. He shows this with a number of case studies that he undertook in Mexico City in 1971. The most striking of these studies is a comparison of the provisional shack of a rag-picker and his family with that of a modern, government-subsidised dwelling lived in by a semi-employed mason and his family.

The dwelling of the rag-picker and his family is near to their source of income, close to family and friends, and cheap enough to allow them to survive with the hope of obtaining a better dwelling as their prospects improve. It thus offers them considerable freedom, and Turner thus refers to it as the 'supportive shack' (1976, p. 54). It is very basic accommodation, yet it fulfils the family's immediate needs and allows them to control their environment.

However, the modern house of the mason's family is located away from their network of friends and, crucially, away from the mason's place of employment. The mason pays out 5 per cent of his income in transport costs to and from work, in addition to the 55 per cent spent on rent and utility charges. Moreover, his wife had previously run a small vending business from their previous dwelling, which was now forbidden under the tenancy regulations. Thus their income has been reduced as their housing and transport costs have risen. Turner refers to this case as the 'oppressive house' (1976, p. 56). Thus, an improvement in material standards can be counterproductive because, being based on abstract standards, they cannot take into account particular needs and conditions. Turner thus concludes from these cases that material standards are not necessarily the most useful measure. He states that 'some of the poorest dwellings, materially speaking, were clearly the best, socially speaking, and some, but not all of the highest standard dwellings, were the most socially aggressive' (1976, p. 52).

Reference

Turner, J. F. C. (1976): *Housing by People: Towards Autonomy in Building Environments*, London, Marion Boyars.

Discussion points

1   Whom do you think social housing is for? Should it be open to everyone?
2   Does it matter who owns social housing?
3   Does social housing still have a purpose?

See also

Chapter 16 Welfare
Chapter 17 Poverty
Chapter 22 Sources of finance
Chapter 25 Housing allowances
Chapter 29 Control

Further reading

King, P. (2006a): *Choice and the End of Social Housing*, London, Institute of Economic Affairs.
Malpass, P. (2005): *Housing and the Welfare State*, Basingstoke, Palgrave.
Power, A. (1993): *From Hovels to High Rise: State Housing in Europe Since 1850*, London, Routledge.

15 Private renting

If owner occupation dominated housing markets at the start of the twenty-first century, the situation at the start of the twentieth century was very different. In many countries it was the private rented sector that dominated, with most households renting. Interestingly, this was not particularly related to social class, with many middle-class households renting. The history of private renting is an interesting case study of how the structures of housing provisions can change without any one particularly intending it. However, before exploring this decline, we need to understand a bit more of what private renting is.

There is no such thing as a typical private landlord and that applies whether we are talking about scale or motivation. At one extreme there are those households who take in a lodger, letting out a room or two. They may be doing this to supplement their income or because they wish to help out a friend. At the other extreme there are large institutional investors with large property portfolios perhaps in more than one country. Perhaps more typical is a landlord with a few dwellings in one area. Many of these owners might actually be using a managing agent rather than manage the properties themselves.

While in some countries private renting has a poor reputation, it actually comprises of a considerable range of property types, from bedsits in multiply occupied houses through to luxury apartments. Some of the most expensive real estate can be privately rented. So, private renting is not just for poor and excluded households.

Indeed, just as it is difficult to suggest that there is a typical private landlord, there is also no such thing as a typical tenant. Clearly many landlords do not operate formal allocations policies or let according to priority need. Their motivation is to ensure a secure income from their properties. This may mean that they take households who need a housing allowance to pay their rent, but this is because it is a consistent income stream rather than because of a social conscience.

The private rented sector is perhaps best suited for transient households, who do not wish to put down permanent roots. This obviously includes students who are only in an area for 3–4 years, but it might also comprise of young professionals who are starting out their careers and more affluent workers who expect to be relocated on a regular basis.

But it is also the case that private renting can be the sector of last resort and so is the tenure for those who cannot gain access anywhere else. There are households who cannot gain access to social housing because they are not deemed to be in priority need, or even because they have a past tenancy experience that excludes them (eviction, rent arrears, etc.). Other households, who are too affluent for social housing but who cannot gain access to mortgage finance, might also find that they have no alternative but to rent privately. Particularly since the 2008 financial crisis, mortgage lenders have sought higher deposits from borrowers and this has effectively excluded many younger households seeking access to owner occupation for the first time. The irony of this situation is that these households are often paying more per month on rent than they would in mortgage payments. Their problem is that they cannot raise sufficient funds for a 20–30 per cent deposit.

An issue that has coloured attitudes towards private renting is whether it is acceptable to make a profit from the housing need of others. Private landlords let properties as a source of income. For some, this means that they are exploiting those in need. In response, we can argue that low-income households are often eligible for state support, with a housing allowance covering all or part of their rent. A more substantive argument is that private landlords are not the only group hoping to make a good living from low-income households. We can make the same claim against supermarkets, shoe shops and clothing manufacturers. The profit motive is actually very common in most walks of life and we can question why housing should be any different. Indeed, those who supply services to social landlords are often private sector companies and the developers of owner-occupied housing are not acting out of charitable motives, either.

However, it is certainly the case that private renting is treated with suspicion by many, and this has led to periodic calls to regulate and restrain the activities of private landlords. The most common call for restraint is over rent levels. It might be claimed that landlords are profiteering from excessive rents, perhaps taking advantage of high demand, and that government should intervene to restrict rent levels. Indeed there is a long history of rent controls in private renting and this is an interesting case study in unintended consequences.

As we have stated, at the start of the twentieth century, private renting was the dominant tenure in many countries. However, a century later, it has been superseded by owner occupation.

There are two reasons that are commonly given for the decline of private renting. One is the provision of large-scale subsidies to alternative forms of provision, particularly social housing, but also tax relief for owner occupiers. Private renting has not had the same level of financial support. Indeed the reverse is the case, and this brings us to the second reason.

Many governments have imposed rent controls on the private rented sector, which means that the landlord may not increase rents above a certain ceiling set by government. This makes rents more affordable and thus within the reach of more households, but it also has the effect of reducing the ability of landlords to make a reasonable return on their investments and so they have tended to leave the sector.

The effect of rent control was to reduce drastically the incentive to supply, because landlords were not able to adjust their rents in line with changes in their costs. However, it also had the effect of increasing demand, because dwellings were let at cheaper rents than would be set in a free market and so were more affordable. This problem of excess demand would ordinarily have been met by an increase in rents, which in turn would have acted as a signal to increase supply. But rent controls not only prevented these market signals from operating, they actually created the opposite effect. Rents could not rise, which was of benefit to existing tenants, but landlords had a reduced incentive to invest in their properties because they could not secure a reasonable return. There was therefore a reduction in the quality of dwellings as repairs and improvements were postponed. In addition, many landlords left the sector, often by selling to the sitting tenant.

The main problem of rent control was that, unlike later forms of subsidy offered to owner occupiers and social landlords, it directly benefitted tenants at the expense of the landlord. Rent control, therefore, is a form of regulation that forces one party to a contract to make a contribution to the other, but without being compensated by government for doing so. Unlike subsidies to social landlords, the government did not provide the subsidy, but rather forced landlords to subsidise tenants by debarring them from increasing their rents. The consequence of this form of subsidy was to present landlords with a huge disincentive to invest or even stay in the market. The policy, however, is cost-free to government, at least in the short to medium term.

A further problem with rent control is that it can affect mobility. If one benefits from a rent-controlled property, then one might be reluctant to leave if one would then have to pay a market rent. This may mean that those benefitting from rent controls might not be the most deserving, in terms of income or need, but just happened to be in the right property at the right time. Thomas Sowell (2007) argues that who benefits and who loses from rent control is arbitrary and dependent on who has the good luck to be inside and the bad luck to be outside. He states that in 2001 a quarter of households in rent-controlled accommodation in San Francisco had incomes in excess of $100,000 per annum.

But, according to Sowell, the problem with rent control is more general. He states that, during the Second World War, low rents caused by rent control allowed young people to move away from their parents earlier and other families to afford larger accommodation than would otherwise have been possible. This created a shortage of rented housing, 'even though there was not any greater physical scarcity of housing relative to the total population' (p. 40). Once rent controls ended after the war, these shortages disappeared. He states that 'As rents rose in a free market, some childless couples living in four-bedroom apartments decided that they would live in two-bedroom apartments and save the difference in the rent' (p. 41). Also young people stayed at home with parents for longer. But, as a result of this change in behaviour, families could now find affordable apartments.

Sowell states that rent controls reduce the incentives for individuals to limit their own use of those scarce resources desired by others. Therefore rent controls can lead to under occupation. He states that in 2001 49 per cent of San Francisco's rent-controlled apartments were occupied by a one-person household, whilst the figure for Manhattan was 48 per cent. Similarly, the elderly have little incentive to vacate rent-controlled apartments that they would normally vacate. Sowell states that 'rent control reduces the rate of housing turnover' (p. 42).

It can be argued that rent controls affect the supply of housing just as they do demand. Sowell states that 'Nine years after the end of World War II, not a single new building had been built in Melbourne, Australia, because of rent control laws there which made buildings unprofitable' (2007, p. 43). Likewise, rent control laws introduced in Santa Monica, California in 1979 saw building permits decline to a tenth of their 1974 level. The reason for this is that investors could not see themselves making a reasonable return on their investment.

But Sowell argues that it is not only the supply of new dwellings that declines, but also the supply of existing dwellings, which landlords neglect to maintain and repair 'since the housing shortage makes it unnecessary for them to maintain the appearance of their premises in order to attract tenants. Thus housing tends to deteriorate faster under rent control and to have fewer replacements when it wears out' (p. 43).

Landlords also do not have to respond to the normal signals, particularly with regard to competition: 'Shortages mean that the seller no longer has to please the buyer. This is why landlords can let maintenance and other services deteriorate under rent control' (Sowell, 2007, p. 51). Because there is a tendency for excess demand in rent-controlled housing markets, landlords do not need to be aware of their competitors but can effectively tell prospective tenants to take it or leave it.

Somewhat perversely, however, the only profitable form of renting is luxury accommodation. This is because this type of rented housing is often exempt from controls. Accordingly, Sowell argues that 'a policy intended to make housing affordable for the poor has had the net effect of shifting resources toward the building of housing that is affordable only by the affluent and rich, since luxury housing is often exempt from rent control' (2007, p. 44).

This is a damning indictment of rent controls. However, the problems are not necessarily solved by abolition. When this occurred in the UK in 1989, the result was a considerable increase in rents, with the effect that a considerable strain was put on Housing Benefit, the UK form of housing allowance paid to low-income households in rented accommodation. As a result, by 2006 the government had introduced limits to benefit payments, and this might be seen as a form of administrative rent control. The consequence of this, however, was that landlords became increasingly reluctant to let to Housing Benefit claimants or would only offer them poorer quality properties.

The history of rent control is interesting in terms of showing the importance of unintended consequences in housing policy. No one particularly sought the decline of private renting, yet, once the process had started, it proved hard to halt. Once policies like statutory rent controls are in place, it is very difficult to abolish them without causing hardship or without an alternative form of support being made available. We might argue that the situation in the UK after the abolition of rent controls was more equitable, in that the government subsidised tenants rather than landlords. However, this did not prevent the accusation that landlords were profiteering from the abolition of controls. Interestingly, land-lords could benefit here without it affecting tenants – who were not paying the rent themselves – and so market signals were still not working. Of course, market signals might only have worked if housing allowances were withdrawn or significantly reduced, but why would we want this to happen?

What this discussion shows is that there is no perfect solution to these problems and that we should be sceptical of those who claim to know how to reform housing. We cannot be fully certain of what will occur as a result of reform, and this perhaps means that we should be careful in how we act. This may not be any comfort to those in severe need now, but then there is no guarantee that what we wish to happen will actually occur.

Discussion points

1   Is it acceptable to make a profit from the housing need of others?
2   Should private rents be controlled?
3   Do we need private renting? Should the state not provide?

See also

Chapter 18 Fairness
Chapter 23 Markets
Chapter 25 Housing allowances
Chapter 32 Reform

Further reading

Albon, R. and Stafford, D. (1987): *Rent Control*, London, Croom Helm.
Dorling, D. (2014): *All That is Solid: How the Great British Housing Disaster Defines Our Times, and What We Can Do About It*, London, Allen Lane.
Sowell, T. (2007): *Basic Economics: A Common Sense Guide to the Economy*, 3rd edn, New York, Basic Books.

Part 4
Welfare

16 Welfare

Welfare is one of those words that get bandied about a lot when discussing housing. It is also a term that has often taken on a specific cultural meaning. For example, if you ask people in both the US and the UK what welfare is, they are likely to say that it means being on benefits: as the phrase goes, 'people are on welfare'. But of course the term has a more general meaning and it is useful to try and separate it from its cultural baggage.

Welfare can be defined in two distinct ways. First, we can suggest that it relates to the well-being, happiness and flourishing of individuals or groups. So in this sense it is a general term that applies in all cases. But, second, we can also define welfare as the means by which this state of well-being, flourishing and happiness is attained or maintained. Hence it can indeed relate to specific financial or material support offered to individuals and groups. So welfare can relate to both an outcome and a means of achieving that outcome.

This suggests that we can see welfare as a quality within individuals – their need to flourish, be happy, and so on – as well as a set of institutional arrangements that seek to create, enhance or maintain that quality. It suggests that resources are necessary for welfare to be attained and maintained, and so we cannot ever fully separate the quality from the means of achieving. These means may well come from the household's own resources. They flourish through using their own income and through caring for and nurturing each other. Welfare in this sense is something internal to the household.

However, welfare is not merely a matter for the individuals concerned. This is because there may be some who, for whatever reason, fail to flourish or maintain themselves adequately. This failure is not only a matter for them, but might have wider societal effects in terms of the spreading of disease, or a higher incidence of crime, as well as impinging on the moral sentiments of other members of society who are concerned that some of their fellow citizens are homeless or in poverty.

So welfare is as much a societal concern as it is an individual one. Hence all states, to a greater or lesser extent, provide some form of financial or material assistance to some, or even all, of their citizens. John Hills (1997) suggests that there are five reasons for the state provision of welfare. First, the state seeks to relieve poverty and redistribute income towards the long-term poor. Second, the state provides social insurance for all against long-term illness, unemployment, early retirement, family breakdown, etc. Third, the state can redistribute income towards particular groups with greater needs, be they medical, familial or related to disability. Fourth, the state can act as a type of savings bank, smoothing out income levels over the life cycle. Households can be taxed and this money used to provide them with residential and health care. Finally, the state can step in where the traditional family structures fail, e.g. divorce and lone parenthood. So states are not just concerned with providing a safety net or dealing with emergencies, but can seek to offer security for all. Welfare provision can go well beyond the basic.

What this definition clarifies is where we might look to sustain our welfare. Hills shows us when the state might need to step in, but his discussion also shows that the state is needed only because other sources of welfare might fail. One key source of welfare is employment and the income that we derive from it. But there may be occasions when we cannot work, particularly towards the end of our lives. Another key source of welfare is the family, but this may break down or be dysfunctional, which in turn might be due to a lack of income. So the state steps in when the family and one's own resources prove insufficient to provide us with the welfare provision we need and expect.

But the level and scale of intervention will differ. Obviously one factor here is the scale of the problem. But also the nature of the welfare good in question needs to be considered. There are some goods that, generally speaking, we are more capable of providing for ourselves than others. This does not mean that state provision might not be needed to distribute this good in some cases. However, some goods, which are fundamental to our welfare, are more amenable to market provision than others, and one of these is housing.

Most European countries have compulsory state-funded education and some form of compulsory social insurance system to provide for health care and retirement. However, their provision of housing is by no means as comprehensive. Nowhere is this more marked than in the UK with its National Health Service (NHS), providing health care free at the point of access, but also with a large owner-occupied sector provided by markets. This suggests that there is something important which influences market activity in housing when compared with health and education. So why is it that a vast majority of housing is provided by markets but only a small minority of health care and education?

The key to answering this question is the distinction between universal and particularist services. Universal services are those that are available to all regardless of the ability to pay. All members of society are eligible for these services. Some of these may be specific to age or circumstance, like free education or pensions, but they are not subject to a means test. In contrast, particularist benefits are targeted at specific groups and are distributed according to a means test. Only those with the relevant circumstances can apply for the benefit and it can be withdrawn if the circumstances of a household improve. In most countries access to social housing is conditioned by a means test, and the same applies to access to housing allowances. Housing as a welfare good, therefore, is nearly always seen as a particularist benefit open only to those with a proven need.

We might suggest that this is a matter of spending priorities and available resources: housing is an expensive commodity and so government naturally targets its resources on those most in need. Yet this answer will not do, because health care is also expensive.

The reason for the differential treatment of housing and health is not really down to money or income but is essentially one of knowledge. In short, we can and do know more about our housing needs than our health care requirements. We can know that we are ill and in pain, but this does not mean that we know the cause of that pain nor the treatment necessary to alleviate it. We therefore have to rely on an expert to diagnose and treat the ailment. Moreover, we can seldom rely on past knowledge to assist us and, even if we could, we would still lack the expertise to treat the problem.

But there is a further problem. The need for health care is contingent on circumstance and so the need is often unpredictable, in that we do not know when or if we are to be ill. All of these issues create very difficult problems for comprehensive market provision. There may be a tendency for there to be underprovision in such systems, particularly amongst the poor who may choose to spend their limited resources elsewhere.

But this is not the case with housing, even though it is an expensive commodity. First, our housing need is permanent, as we will always need housing regardless of our circumstances. What differs, of course, is whether it is currently fulfilled or not. Second, because we always need housing, this means that it is predictable, allowing for a more regular pattern of provision. Of course, our need may change as we start a family and become more affluent, and then downsize as we get older, but this is not often due to any sudden change that demands an immediate intervention like an emergency operation. Barring war and natural disaster, there is therefore the ability to plan any change in a predictable manner. One other way of looking at this is to suggest that the stakes are often lower with housing than with health care. Poor quality housing may indeed be serious and need sorting out as quickly as possible, but poor health or a sudden illness clearly requires a more rapid response.

Third, as housing is both permanent and predictable, with a slower and more regular pattern of change, it is more readily understandable, in that we know we need it, that we will always need it and to what standard we require it. Even when we ask homeless people, they are entirely capable of telling us what constitutes good housing and they can recognise it when they see it. Unlike the case of health care, we do not require an expert to tell us that the housing is good and fits our needs.

These three principles – permanence, predictability and understandability – suggest that decision-making in housing can be devolved more readily to the level of the household, and thus housing is more amenable to choice. This does not mean that we can build or maintain our dwelling ourselves (although we might), but that we have sufficient knowledge to set the parameters and determine what we need.

So we can suggest that housing is more amenable to choice within markets. The nature of housing, being predictable, permanent and understandable, makes it compatible with individualised decision-making in a way that is perhaps not possible with other complex welfare goods. However, this does not mean that housing markets operate perfectly. Indeed, they may work rather badly from time to time, especially if we take meeting everybody's housing need as the key measure. However, housing markets clearly do operate and can be said to work in that most people are well housed most of the time. But governments quite naturally need to concentrate on those who are not well housed, and this means that they need to intervene in housing markets in a targeted manner, supporting those households who are not capable of supporting themselves.

One of the consequences of this dichotomy between universal and particularist welfare is that the former tends to be rather more popular than the latter. Those services that are open to all tend to garner public support and this, in turn, means that politicians are keen to promote their commitment to universal provision. In addition, it is easier to justify universal provision: we are all paying for it and we all get something back. It can be seen as a form of social solidarity and provide a sense of identity. We can see this particularly with the National Health Service in the UK. However, politicians have less incentive to support particularist benefits that are only used by a minority. Moreover, this minority might not currently be contributing through the tax system. The result is the tendency, often played on by politicians, to create an 'us and them' situation between those paying for a service and those receiving it.

There is, then, a clear political dimension to welfare, and we need to be aware from whence this derives. While we can point the finger at politicians pandering to owner-occupiers, it is also clear that there are some clear reasons for the relative position of the housing tenures, and it is unlikely therefore that this can be unravelled simply by a change of policy.

Discussion points

1   Who is responsible for the welfare of your family and yourself?
2   Is there a real difference between health and housing in terms of welfare?
3   Should we see social housing as a safety net or does it have a wider purpose?

See also

Chapter 7 Need
Chapter 8 Choice
Chapter 14 Social housing
Chapter 18 Fairness
Chapter 30 Government

Further reading

Hills, J. (1997): *The Future of Welfare: A Guide to the Debate*, revised edn, York, Joseph Rowntree Foundation.

King, P. (2003): *A Social Philosophy of Housing*, Aldershot, Ashgate.

King, P. and Oxley, M. (2000): *Housing: Who Decides?*, Basingstoke, Palgrave Macmillan.

Levine, D. (1995): *Wealth and Freedom: An Introduction to Political Economy*, Cambridge, Cambridge University Press.

17 Poverty

Historically speaking, it would be hard to argue with the premise that the principal justification for state intervention has been the relief of poverty. Whether we talk in terms of need, social justice, rights or any other concept, the basis of the problem that we are trying to solve is that some households lack the ability to pay for good quality housing, as well as most other goods. Being in poverty prevents people from flourishing, and so their needs remain unmet, resources are distributed unjustly and the rights of some citizens are not fulfilled.

As it says in the Bible, the poor are always with us, and indeed poverty has been a perennial problem. Even as societies become more affluent, there is still the problem of poverty. This might be because some people miss out, or it might be because economic development is not evenly distributed across the globe. So, while a society or even the world might be becoming wealthier, there may be significant pockets of poverty remaining.

But the quote from the Bible is not as straightforward as it looks. The point is not that the author was being callous, even though they might be being realistic. What they are actually doing is not just reminding us that there has always been poverty, but that the poor are with us. We cannot separate ourselves from the poor and ignore them. This, then, is an injunction to recognise that poverty exists and that we must do something about it, because the poor are part of us. The poor cannot be left on their own, forgotten about and ignored; they are part of our community and so we have a moral responsibility to deal with poverty.

One of the reasons why this issue of poverty causes debate is the manner in which it is defined, particular by policymakers. We can argue that the general public, little concerned with the niceties of policymaking and not really having thought terribly hard about the problem, might see poverty as a severe lack of money. However, most definitions of poverty are not this simple, being based instead on an understanding of the relative position between those at the bottom and the average for that society.

The simplest way of defining poverty is to determine a minimum level necessary to meet basic needs – the poverty line – and, if a person's income falls below this line, then they are in poverty. We might wish to set this at a global, national or local level, and it will doubtless change as incomes rise and fall over time. This might mean that the poverty line in sub-Saharan Africa would be different from the US and Canada, and the poverty line in Canada would be different a 100 years ago compared with now.

The problem of this definition is that it is dependent on finding some consensus on what constitutes basic needs. How do we decide what is basic and essential and what is non-essential? We might wish to come up with a list of essentials, but would these apply in all cases and for all time? In addition, if we are concerned with human flourishing, we do not just want to keep people alive but we wish to see them achieve their full potential and be able to participate fully in society.

As a result of the problems of maintaining an absolute definition, an alternative has been developed. This is what is referred to as relative poverty. This is where poverty is defined relative to the standards of living in a society at a specific time. People are in poverty if they do not have a sufficient income to meet their needs and are excluded from taking part in those activities that are deemed to constitute a normal life in that society and in that time. So the concern goes beyond the basics to consider what is an acceptable life in, say, Canada in 2015. We have to see what is taken as a normal life and this might include Internet access, holidays, TV and a computer.

In practice the distinction between absolute and relative definitions is not complete, because definitions of relative poverty will also fix a line below which persons are deemed to be in poverty. However, this line will not be expressed in terms of a basket of goods that fulfil basic needs, but as a percentage of average earnings. As an example, the successive UK governments have used 60 per cent as the measure. Thus a family receiving less than 60 per cent of average earnings is in poverty. Clearly, in a country like the UK, this would not be considered absolute poverty (although remember that some households may be well below the 60 per cent threshold).

The notion of relative poverty, while it has the advantage of shifting as incomes change, is a contested one, and some have gone so far as to suggest that it is logically absurd. For example, it can be argued that relative poverty could be substantially alleviated by making a sufficient number of those above the threshold worse off. Indeed, several countries noted a fall in poverty after the 2008 financial crisis, but this was entirely due to a general drop in earnings linked to the cushioning effect of welfare benefits for those at the bottom. In other words, many in work saw their living standards reduce, while the level of benefit payments either remained stable or increased with inflation. It would be perverse though to trumpet this fall in poverty as beneficial to anyone.

Another fault in logic often pointed out is that relative poverty can increase even in a situation where every household's income is increasing. This can occur where the increase for those on high incomes is greater than those on medium or low incomes. This might mean that average earnings increase faster than the increase of those on lower incomes, and so poverty appears to increase. Indeed, in a society where everyone is a millionaire there could still be relative poverty depending on the distribution of income.

The argument, therefore, is whether these logical problems are fatal to the concept of relative poverty. Most commentators would argue that they are not, and not merely because they are rather improbable. Of course, a philosopher might say that it is the principle that counts here, but in the real world it might be difficult to come up with a more workable measure. Indeed, a more serious criticism of relative poverty as a measure is that the percentage of average earnings is set too high. This, though, is a political issue rather than one of principle and can only be settled by debate and discussion around what it actually means to participate in society.

What is clear, however, is that the concept of poverty has become inextricably linked with another substantive concept, namely, inequality. The issue that lies behind the concept of relative poverty is not that individuals are actually starving and on the streets (although this might be the case), but the relative position between those at the top and those at the bottom. This has led critics of the relative definition to see it as little more than a cover for campaigns against inequality.

A further debate surrounding poverty concerns what causes it. In essence we can reduce this debate to two explanations: individual behaviour or social structure. On the one hand there is the argument that poverty is caused by simple bad luck or by the poor life choices of the individuals concerned. They may have behaved badly, or not be prepared to work, or have taken some decision that means that they cannot participate fully in society. In response to this, society may first wish to decide whether the person merits help, what in the nineteenth century might have involved separating the deserving from the undeserving: the widow and orphans from the dissolute gin queen. Once this distinction had been made, the deserving could be assisted by charity or by state provision on the basis that increasing their income would deal with their problems. The widow would get a pension and her problem would be solved.

On the other hand, however, is the argument that poverty is caused by broader social factors that are beyond the competence of any one individual or group in society to deal with. Poverty is not due to any failing on the part of the individual, but is due to issues such as racial discrimination, sexual and gender discrimination, and the ingrained effects of social class on educational achievement. This means that an individual is powerless to deal with their poverty and the problem can only be solved by a fundamental reorganisation of society. This can only be done through the agency of government, which can seek to change the structure of a society.

We can argue that this structural view of poverty is the one that has had the greatest effect over the last 50 years in many countries. Hence, in the US in the 1960s, the issue of poverty was very much linked to civil rights and eradicating racial discrimination. The problem could only be dealt with through state intervention by the passing of civil rights legislation and the introduction of welfare benefits and the provision of services. This view of the problem, and the remedies used, were mirrored in many European countries, which saw a rapid increase in welfare expenditure from the 1960s onwards.

This structural view has in recent years come under pressure from those suggesting that it has been a mistake to ignore human behaviour. Some critics of welfare policies in the US, such as Charles Murray, have asked why it is that government spending on poverty relief has increased even as American society as a whole has become more affluent. His argument, which it is fair to say is by no means universally accepted, is that many individuals changed their behaviour in response to the incentives presented by government programmes. While in the 1950s there was stigma attached to certain activities, such as lone parenthood, these now attracted welfare benefits. These welfare payments were means tested, which meant that higher payments went to those with the most severe problems. This, according to Murray, meant that there was no incentive to better oneself, and that people deliberately made their situation worse.

Murray's argument is that government action can actually influence poverty, but unfortunately not in the manner that government would hope. Instead of improving the situation, it merely institutionalises poverty and makes it worse, and this is because individuals alter their behaviour according to the situation that faces them. While Murray's views are still considered controversial, it is fair to say that there has been something of a comeback for behavioural arguments in relation to poverty. It is now much more common for politicians and policy-makers to accept that government actions are not neutral and that it is therefore important to understand the effect that incentives play on individual behaviour. We can see this view enacted in policies across the world such as attaching conditions to welfare benefits based on an individual's job-seeking behaviour and the introduction of fixed terms to benefit entitlements and social housing tenancies. The idea behind these policies is that they encourage the 'right' sort of behaviour, such as gaining and sustaining employment.

But these policies do not involve ignoring poverty. They are merely another means of tackling a seemingly intractable problem. So we might conclude that, if the poor are always with us, so are our attempts to deal with poverty. This is one issue where failure does not mean that we stop trying.

Think piece: Relative poverty

The main way in which poverty is measured in developed countries is by using a relative measure. So a household earning less than a certain percentage of average earnings can be in poverty. But does this make sense?

- A household's income could be increasing, but still fall below the average. How can this make them poor?
- In a society of millionaires, those with the least income might be deemed poor. Should we bother about this?
- Are we really concerned with how people can live and whether they can participate fully in society?
- What does it mean to participate fully? Why should this relate to the goods and services that I can buy?
- Why don't we use an absolute measure of poverty?

Discussion points

1   Should we see poverty in absolute or relative terms?
2   Can you be poor and own a 42-inch plasma television?
3   Is poverty caused by such things as inequality, class and racism, or by individual behaviour?

See also

Chapter 6 Social justice
Chapter 7 Need
Chapter 14 Social housing
Chapter 19 Inequality
Chapter 21 Crisis

Further reading

Dorling, D. (2014): *All That is Solid: How the Great British Housing Disaster Defines Our Times, and What We Can Do About It*, London, Allen Lane.

Murray, C. (1985): *Losing Ground: American Social Policy, 1950–1980*, New York, Basic Books.

Niemietz, K. (2011): *A New Understanding of Poverty*, London, Institute of Economic Affairs.

Townsend, P. (1979): *Poverty in the United Kingdom*, London, Penguin.

18 Fairness

Most of us have an intuitive sense of fairness. As children, we would often claim that it was not fair if we were not included in a game or did not get an equal share of cake. These are perhaps some of the strongest feelings we have and, it seems, no one needs to teach us about fairness. However, fairness is one of those concepts that is used in many different ways and can be used to justify or criticise a lot of initiatives.

There has been a lot of criticism of housing markets since the financial crisis in 2008. One of the contributory factors of the crash was the sale of so-called sub-prime mortgages to households with low incomes and poor credit histories. They were sold these mortgages on the basis of low initial payments, and there is some evidence to suggest that the full implications of these mortgages, particularly that they were variable rate and interest only, were not explained to those buying them. In effect, financial institutions were seeking to benefit from selling dubious products to households who had little room to manoeuvre once the market started to decline. This, it is argued, was unfair, particularly in the sense that it was the households who lost everything, while the bankers were often protected and are still in business. This situation is what is often described as an unfair choice. The households in question did not, formally speaking, have to take out a sub-prime mortgage. Indeed, no one is forced to do this. However, it might be argued that they were faced with no other options, so, if they wished to buy a house (in a country with relatively little subsidised housing), they had no other choice.

A different example of an unfair choice relates to the change in payment of Housing Benefit in the UK. A peculiarity of the Housing Benefit system was that all local authority tenants had their benefit paid direct to their landlord. In the housing association and private rented sectors it was also possible to request that the payments were made direct to the landlord, and this became the default option for most landlords, with tenants doubtless seeing it as the most convenient option. However, it is now government policy that all welfare payments, including Housing Benefit, are paid directly to the tenant. It is argued that this will make tenants more responsible in that they would now have to budget in the same manner as a working household. This, it is suggested, will ease the transition from benefits into work. However, on the negative side, because their rent is now paid to them, rather than their landlord, some claimants have found that their income has effectively doubled or even trebled. These households have never had to actively pay their rent before and may be tempted to use this apparent increased income for other purposes. Of course, we can argue that these households are under a legal obligation to pay their rent, and risk eviction if they do not, but being on a low income presents them with the difficulty of doing the right thing in the face of competing financial pressures. So, it can be argued that, instead of making these households independent and responsible, they are actually being set up to fail.

There are a number of examples of apparent unfairness that relate to owning housing. Perhaps the most fundamental point is whether it is ever fair for someone to make a profit fulfilling someone else's basic need. Is it acceptable for a landlord to charge a market rent to a household that would otherwise be homeless? It might be argued that this household can claim support from the state, but this does not really answer the complaint that the landlord is profiting from the misfortune of others. The obvious answer to this complaint is that many other necessities, such as fuel, food and clothing, are also provided by private companies who expect to make a profit. One can also question in what other way housing might be provided without the profit motive.

Second, is the question of how it can be acceptable for some people to live in a huge mansion while others are on the streets. Why is it acceptable for some to live in luxury and excess while others are dying from exposure? This, of course, goes beyond housing and is really a question of why poverty is ever allowed to exist in an affluent society. In other words, how can we justify inequality? A related issue is that some households, who are already relatively housing rich (being established owner occupiers), will inherit property from their parents and will expect, in turn, to pass this on to their children. However, those in rented housing have no assets to pass on. In this sense, housing is a key source of inequality and this has led some to argue that large dwellings should be taxed. This already occurs in some countries that have inheritance taxes, but there is also an argument for a recurring tax on high-value properties. One argument against this so-called 'mansion tax' is that it is based on the hypothetical value of the property, rather than the income of the household and therefore their ability to pay. Some households, particularly the elderly, may be asset rich, but income poor, and so this tax, justified in terms of fairness, can be itself considered unfair.

Another example of possible unfairness with regard to housing markets relates to the fact that households living next door to each other can be paying vastly different housing costs. One's housing costs depend on the value of the property when it was purchased, and with house price inflation this will mean that someone who bought a house 25 years ago will be paying only a fraction of what their neighbour did who bought the property a year ago.

There is now a different sense of fairness to consider that relates to the nature of housing. In societies where social housing or housing allowances are provided, there is effectively a situation where one set of households pay for the benefit, while others, who might be paying no tax because they are not working, receive it. If social housing is allocated to those in need, then these will mostly be households on low incomes who pay less tax. Indeed, it may well be that these households are not working and receive a housing allowance to help them pay their rent. But social housing is only available to those in severe need, and housing allowances are paid only to those on low or no income. Therefore those who are paying for the service are not eligible to access any of its benefits. This can be perceived as unfair and might lead to resentment on the part of those taxpayers who see themselves as excluded. Having said this, others may argue that it is in the very nature of a civilised society for those with a sufficient income to support those who are destitute and in need. This is a valid argument, but it does tend to ignore the fact that tax systems are not voluntary and there is no facility to opt out.

This argument about relative inputs and outputs is relevant in that some governments have capped their welfare systems, limiting total out-of-work benefits to the level of average earnings. Interestingly, politicians in the UK, for example, have justified this capping on the grounds of fairness. However, it is not fairness with regard to the least well off but to those working on average earnings that forms the basis of this argument.

But, we might ask, why do most developed societies have welfare systems, including housing allowances or subsidised housing, if it is not because of the perceived unfairness of some households who can afford to be well housed while others can afford nothing at all? So, after considering these examples, we might spend some time considering just what fairness is. What ought to be clear, however, is that fairness is not a straightforward concept and that it can be applied in many different and even contradictory ways.

We can define fairness is a number of ways. We can suggest that fairness is where we are free from bias or favouritism. It is where we do not prejudge an issue or start from any particular presumption. In this sense, the opposite of fairness is prejudice, which is precisely where we come to a situation having already arrived at a judgement regarding that situation. What we lack when we are prejudiced, therefore, is the requisite sense of impartiality, and instead have already taken one particular side. Accordingly, no evidence can sway us, as we have already made up our mind. Fairness is therefore where we are able to take a balanced view based on the facts, reason and common sense. We may start with a particular view but, by listening to a reasoned argument, we can be persuaded that our original position was mistaken and accordingly shift our position to accord with what reason now dictates. Fairness can also refer to consistency, in the sense of adherence to rules, logic or ethics. This would be when the same rules or principles are used in all cases so that all claimants or applicants are treated the same. The term also carries with it the sense of being lawful and proper, in that procedures have been followed.

Perhaps the two most common ways in which fairness is manifested are in regard to either equality or merit. Fairness means that all people are treated equally. This relates to the intuitive sense of fairness. However, it might not always appear to be fair. For example, if we pay everyone the same regardless of the work they do, this might be fair in the sense that all people have the same income and so can all meet their basic needs. However, those who have rare talents, who work especially hard or have difficult and dangerous jobs might not consider it fair that they are paid the same as those who have much easier jobs. So, it might be that those people with special merit, such as talents and skills that they have honed over a long period of education and training, should be paid a higher wage. Again, this appears to link to the intuitive sense of being more deserving and having earned a higher reward.

So both equality and merit can be seen as fair, but clearly they will conflict. Both of these conceptions of fairness relate to proportionality and can be justified in this way, but clearly we cannot distribute income equally and according to merit at the same time. We therefore would need to make some form of judgement over which form of fairness we wish to use in allocating resources. This means, however, that any attempt to make a fair distribution of resources will be both controversial and contested.

A further sense of fairness which we need to consider is that of reciprocity. This, too, relates to proportionality, but it is a rather more overt concern for the relative position between claimants. It is where we look at how much we have contributed compared with how much we receive and then compare this with others. We then determine whether the package of benefits and burdens that we receive is fair. This sense of fairness does not directly relate to the level of benefits, but rather whether we are getting what we believe we deserve compared with others. Finally, we can see fairness as being a concern for how decisions have been arrived at. We might be happy with an adverse outcome, so long as we are assured that the process was done fairly and without bias.

What ought to be clear is that all of the examples of fairness discussed at the start of this chapter are legitimate. However, they cannot all reside together. This suggests that one particular sense of fairness might be dominant in a particular situation. The end point of this, however, might be that an outcome that seems eminently fair to some is considered outrageously unfair by others. So what matters is who is taking the decision and what ability others have to challenge that decision. This suggests that, at a practical level, perhaps the most important sense of fairness is that of procedure, and so we should be concerned first with how decisions are taken and only then with the outcomes.

Discussion points

1   Is it fair to help some people and not others? How do we decide if it is fair?
2   Is there such a thing as an unfair choice?
3   How does equality differ from fairness?

See also

Chapter 6 Social justice
Chapter 16 Welfare

Further reading

Elster, J. (1995): *The Cement of Society: A Study in Social Order*, Cambridge, Cambridge University Press.
Rawls, J. (2003): *Justice as Fairness: A Restatement*, Cambridge, MA, Harvard University Press.

19 Inequality

It is an interesting fact that, when many people talk about the problems of poverty, or when they discuss the unfairness of a government policy, or the class system, or the elitist nature of higher education, what is actually being objected to is inequality. What really concerns many people is the disparity between those at the top and those at the bottom. Accordingly, the key aim of many is greater equality, leading to the ideal of an equal society.

What makes this important is the widely held perception that, since the 1980s many societies appear to have become less equal. The differences between those at the top, in terms of their lifestyle, housing and income, is so very different from what is experienced not just by those at the bottom but by those on average incomes. Inequality, it appears, is getting worse, and those at the top seem to be protected. For example, those in banking and financial services paid themselves fabulous salaries before the financial crisis in 2008, then insisted that governments bail them out and seemingly carried on paying themselves fabulous salaries, even as the rest of us saw our living standards fall and our taxes increase. And then it was found that many of those working in financial services could afford to pay accountants and lawyers to help them avoid tax.

Since 2008 there has been a renewed discussion on inequality, most notably following the publication of *The Spirit Level* by Richard Wilkinson and Kate Pickett in 2009. This book purported to show that equality worked to the benefit of all and that those societies that were more equal were happier and had higher levels of well-being. The authors claimed to show that a myriad of social problems including obesity and teenage pregnancy were lower in societies that were more equal. They went so far as to argue that even the wealthy benefitted from greater equality. Not surprisingly this has proven to be a controversial thesis that has led many to either seek to debunk or reinforce Wilkinson and Pickett's position.

What it does show is that the issue of inequality matters to a lot of people. But what are we actually talking about here? We often talk about inequality, but inequality 'of what?' This is because we quite clearly tolerate some forms of inequality. We expect sporting teams to contain the very best representatives and not an average sample of the population. The eleven players who turn out to represent England at football are an elite. They are hopefully the very best footballers in the country, and England supporters would accept nothing less. In schools and universities we give different grades to students based on their intelligence and their efforts. We acknowledge and reward genius in the sciences, music and the arts. So we recognise different levels of talent and ability, even as we might be jealous of it, and we accept competition in some areas even as we know that the result will be inequality.

But the problem of inequality is not whether we are good at sports or get a lower mark than someone else in our exams. There may well be some who disagree with competitive sports and think that all must have prizes. But the real issue for those concerned with inequality is economics. What concerns them is the distribution of income and the consequences that flow from this. So the concern for inequality is over the life chances of those on low incomes compared with the wealthy. In terms of housing the concern is for access and affordability. Can low-income households gain access to good quality affordable housing? Are some households excluded from housing markets, and how can this be alleviated? This becomes even more of an issue in high demand areas where many households are priced out of the market.

One way of dealing with housing inequality is to tax high-value properties. The justification for this is that these are likely to be high-income households or that they have benefitted from a considerable appreciation in the asset value of their property. The income derived from this 'mansion tax' could then be used to fund housing for those on a low income or for other equally beneficial expenditure. However, there is a problem with this form of taxation, in that there is no necessary correlation between a high income and property value. It might well be the case that some households in high demand areas, particularly the elderly, might be asset rich but income poor. Having retired, they now subsist on a relatively low and fixed income, yet have a high-value property that they may have bought up to 40 years earlier.

If inequality is such a bad thing, would getting rid of the rich deal with the problem? If rich people did not exist, would we all be happier as a result? This might lend itself to an easy political slogan, but does it actually make sense? It is doubtless much simpler to level down rather than to raise up. It is easier to confiscate from the wealthy than to ensure that the poor can properly compete with them. But would this be a sensible approach? It is noticeable that Rawls, whose theory of justice is precisely concerned with benefitting those at the bottom, recognises that we respond to incentives, and so some inequality will be necessary if we are going to provide the resources to help those in poverty. We need doctors, entrepreneurs and those who create jobs, and so we have to provide the right incentives.

But, also, we can argue that we can only achieve equality by constantly interfering in the decisions that individuals make. We spend our income how we choose. Some of us invest wisely and others spend frivolously, and the result is that some end up with more money than others. This has not come about by exploitation or illegal means but accumulated voluntary actions. So, why should the outcome be seen as problematic, even it means that one person is wealthier than another? What this suggests is that equality could only be achieved through continual intervention to redistribute income even though the unequal outcomes were the result of voluntary activity. This situation is demonstrated in Robert Nozick's famous 'Walt Chamberlain' thought experiment in *Anarchy, State and Utopia* (1974). Nozick asks us to imagine a community in which income equality has been established. This community likes basketball and their team boasts the most talented basketball player, Walt Chamberlain. Such are his talents that many individuals choose to pay a voluntary premium to watch Chamberlain, which is then paid to the player. Accordingly, at the end of the season Chamberlain is much wealthier than anyone else in the community. But this has arisen entirely from voluntary acts and without any coercion. Accordingly, Nozick questions on what basis it would be acceptable to return to strict equality, a situation that could only be arrived at by considerable and continual intervention – taking the money from Chamberlain and returning it to his fans – when the inequality had derived entirely from voluntary acts by people enjoying themselves. This leads Nozick to suggest that, if we wish to be free to live how we choose, then we have to accept inequality and if we see equality as all-important, this will lead to a considerable restriction in individual freedom.

What this suggests is that we cannot avoid a degree of inequality. It is hard to see how making us all worse off would make us happier, and if that means that some of us are better off than others, then it would perhaps be seen as a price worth paying (after all, we all think we will be the ones who will succeed). The problem is how to ensure that it does not get out of hand in a globalised world where the wealthy (which probably does not include us) are footloose and can take their money with them.

Discussion points

1   Would a society where everyone earned the same, regardless of their talents, be a society you would like to live in?
2   Does some inequality make us better off?
3   Should we tax those who own expensive houses?

See also

Further reading

Dorling, D. (2014): *All That is Solid: How the Great British Housing Disaster Defines Our Times, and What We Can Do About It*, London, Allen Lane.

Nozick, R. (1974): *Anarchy, State and Utopia*, Oxford, Blackwell.

Rawls, J. (1971): *A Theory of Justice*, Oxford, Oxford University Press.

Wilkinson, R. and Pickett, K. (2009): *The Spirit Level: Why More Equal Societies Almost Always Do Better*, London, Allen Lane.

20 Homelessness

Homelessness is almost certainly one of the most desperate situations we could expect to face. Not having a roof over our head puts our health in jeopardy, makes it hard for us to get or keep a job, puts a huge strain on relationships and makes it impossible to participate in education and to ensure our physical security. We should note in this regard that, when we discuss this problem, it is not referred to as 'houselessness'. Of course, what is lacking is housing, but what really matters here is what we are able to do once we have a suitable dwelling. What we lack, therefore, is a home that allows us to have a secure and private life and to share this with those whom we choose, rather than being without the ability to exclude ourselves from the world.

Much of what we take to be a normal life is simply not possible without a home and housing. It is therefore not surprising that dealing with homelessness is often a priority in many housing systems. The homeless are given priority, and many statutory and voluntary organisations are involved in assisting them, whether it be providing food and blankets, running hostels and night shelters, lobbying for more resources, right the way through to direct provision.

But the connection with home also suggests that we may have a problem even if we technically have a roof over our heads. Therefore, in some countries, such as the UK, homelessness is defined in statute not merely as not having housing, but where there is a lack of secure, sustainable and safe accommodation. It is not just important that we have somewhere to live, but this has to be safe and it has to be reasonably permanent. Thus, a woman suffering from domestic violence may be considered homeless even if she has her name on the lease or the deeds. What matters is whether she can hope to reside there in safety into the future or if the threat and actuality of violence makes this impossible.

What this means, however, is that countries will have differing definitions of homelessness. It also means that, in some countries, people can actually be living on the street and not be seen as homeless because they do not fall into the categories determined by statute (this often applies to single people). So, for example, in the UK there has tended to be a distinction made between those who are statutorily homeless and those who are roofless.

The essence of homelessness is found in the concept of exclusion. On the one hand the homeless person is excluded from all private property and perhaps much public property as well. There is no place where they have the right to be. As a result, they are also excluded from much of society, unable to claim benefits, register with a doctor, pursue education or get a job. But we might suggest that the homeless are also unable to exclude. When we go home at night we can lock the door and close ourselves off. We only allow those into our dwelling whom we choose, and so we exclude all unwanted others. But the homeless, sleeping on a park bench, a shop doorway or even in a hostel, might not be able to do this. They have no means, legal or physical, to exclude others. They have no privacy and no security. Everything they do is shared and in public. They are not able to put a boundary around themselves like the rest of us when we lock the door and draw the curtains.

Having said all this, the priority provision of housing to the homeless is not without controversy. It has been argued, for instance, that prioritising certain forms of provision merely leads to the expansion of that needs group. This is an argument used by critics of state welfare such as Charles Murray who has wondered why state-funded poverty relief has increased even as society became more affluent. His argument, itself hotly contested, is that state provision gives households an incentive to change their behaviour to make them eligible. The argument is then that some people might 'make' themselves homeless because this is the fastest route into social housing. The problem with this argument, however, is that it almost always rests on anecdotal evidence rather than anything more concrete. Indeed, even if we can verify that a number of people behave in this manner, it is still a leap to suggest that this behaviour is in anyway general.

What gives credence to these sorts of arguments is scarcity. There is hardly ever enough good quality housing to go round and this means that there needs to be some form of allocating these scarce resources. One means of doing this is to allocate according to need, such that those who have the most pressing need are prioritised. This will inevitably mean that homelessness, however it is defined, is prioritised, and this will be at the expense of other households, all of whom might claim to have a legitimate call on the state's resources.

There is a link here with the concept of responsibility. If we believe that the homeless should be prioritised, we might not worry too much about how households came to be in this situation. However, others may worry that these households have got into this situation because of some act or omission for which they should be held responsible. They might have been evicted due to rent arrears or because they harassed a neighbour. Why should these house-holds be prioritised over and above other households who patiently wait their turn on a housing list, who pay their bills and treat their neighbours with respect?

These are indeed contentious issues and they do not lend themselves to easy and convenient answers. However, they do point to a number of problems that are central to any housing system. First, is the perennial problem of scarcity: as long as we have limited resources these questions will continue to be asked. Second, they question what the role of the state is and how far it should go in providing for its citizens. And third, there is the question of what individuals should be expected to do for themselves. Should they expect state assistance regardless of their actions or should we make support conditional on their behaviour? One thing is certain: as long as there are homeless people these questions will persist.

Discussion points

1   How important is the ability to exclude?
2   How do we decide who is the main priority for assistance?
3   Should we limit assistance only to those who deserve it? And how do we decide this?

See also

Chapter 10 Responsibility
Chapter 14 Social housing
Chapter 17 Poverty
Chapter 18 Fairness

Further reading

Daly, G. (1996): *Homeless: Policies, Strategies and Lives on the Street*, London, Routledge.

McNaughton, C. (2008): *Transitions Through Homelessness: Lives on the Edge*, Basingstoke, Palgrave.

Murray, C. (1996): *Charles Murray and the Underclass: The Developing Debate*, London, Institute of Economic Affairs.

21 Crisis

If we listen to many people talking about housing, there seems to be a permanent housing crisis. Indeed the use of the word 'crisis' is now almost obligatory. It is not enough to say that things are a bit difficult or that the government is not doing a good job. Instead, it must be a crisis. Clearly, there is a political element to this, but the term is now so ubiquitous that it often goes unchallenged. Without any sense of irony, the fact that housing is in crisis is now taken for granted.

But what is a crisis? A fairly standard dictionary definition informs us that the root of the term is medical and refers to a decisive point when action has to be taken. More generally, a crisis is a period of intense difficulty or change. It is the point when an important decision must be taken. We are at a turning point and can no longer delay taking action. So we can say that a crisis is an unsustainable situation that needs immediate attention. This being so, how can we have a permanent crisis? A crisis is something that is unsustainable and cannot be contained. It must be dealt with now or else.

Now it might be that we are unfortunate enough to be experiencing a series of different crises one after the other. But this would seem to be rather unlikely. In any case, the term is always used in the singular: it is just the one crisis. Hence we can legitimately question why the term is used with such apparent abandon.

The most obvious reason to claim a housing crisis is as part of a claim for resources. In times of restricted public spending – and when is there not a limit on this? – no one succeeds in winning much attention by claiming that things are 'not that good' or 'could be better'. To get the resources they believe are necessary – and there is no need to doubt anyone's sincerity here – they must shout louder than others making equally valid claims (with the same degree of sincerity). If they are to win the resources they feel are urgently needed, then there is surely no harm in overstating the problem somewhat. After all, this is what everyone else does.

The fact that it is a claim for resources becomes clearer when we realise that the shouting is being done by those with a vested interest in the sector, in the form of housing professionals, academics, commentators and activists. The case for the housing crisis is not being made by the general public, or by those holding the resources. Not everyone sees housing as the top priority, and those who claim that there is a crisis are the relatively small number of people who do.

But, of course, the fact that it is a relatively small number of people means that they need to shout even louder to be heard, and thus it is not hard to see why the rhetoric is raised even further.

We have to appreciate that, with a few really severe exceptions, any problem has to be viewed relative to others. So, is the problem in housing as severe as that in health or education? Which problem needs the resources the most? Of course, this is not always easy to answer, depending as it does so much on how the nature of the problem is perceived and communicated and what we can use to bolster support for our claims. Perhaps then, we might come to the perverse conclusion that the reason that there are so many calls for a crisis is precisely because housing is commonly seen as less important than health, education and other areas of public policy.

Of course, the situation even in the most developed countries may be far from perfect. There will be people on low incomes who live in poor housing conditions, and we should not diminish the impact of this on those concerned, nor should we be sanguine about what the fact of even one person living in poor housing says about the society that we live in. But the situation as a whole is not, properly speaking, one of crisis.

So what would constitute a genuine crisis? We might say that an environmental disaster such as an exploding volcano, earthquake or flood that destroys a large number of dwellings would be properly called a crisis, as would a famine leaving people destitute and unable to feed themselves. We might also see a housing crisis as one of the consequences of war, which often leads to physical destruction and the mass displacement of population, leading to refugees living in makeshift accommodation without necessary sanitation, water and other basic supplies. A crisis might have an economic cause such as collapse in a nation's financial system, which leads to mass unemployment, homelessness and perhaps the inability of government to meet its internal and external commitments.

All of these events are on a large scale, but there can be personal crises due to a whole range of issues such as relationship breakdown, loss of employment, violence, fire, theft, addiction and so on. These are, of course, the most common crises, happening to someone, somewhere, every day. These crises need immediate attention and should never be dismissed or diminished. Yet these are not endemic to the system. They have not been caused by housing systems, but rather are the very reason why government intervenes, and they still persist despite the existence of support. These crises arise from a myriad of causes and are largely due to specific individual issues that no system could pre-empt. It is only possible to react to many of these personal crises. Those arguing that there is a housing crisis may suggest that it is merely the summation of these individual crises, but this would necessitate some general cause and so we require an explanation of why an apparently general condition only affected certain individuals and not others.

But does it really matter if the rhetoric gets a bit out of hand, especially if it might work and leads to a greater allocation of resources for housing? I would suggest that there is indeed a very real problem in crying wolf too often, and that is that one will no longer be taken seriously, but instead be ignored as someone who can no longer be trusted. In the competition for scarce resources it becomes too easy to simply suggest that the housing lobby is frivolous, and so what do we do when there is a genuine crisis and we find that no one is listening to us any more?

Think piece: Montserrat – a real crisis

Montserrat is one of the Leeward Islands in the Caribbean. It is a small island only 16 km long and 11 km wide. Its population in 2015 was just over 5,000 people. However, in 1995 the population was over 13,000.

In July of that year a previously dormant volcano, Soufrière Hills, erupted, destroying part of the capital city. The volcano has affected around 90 per cent of the housing on the island and caused an estimated 8,000 people to flee, most of whom have not been able to return. The island's airport has been buried under lava and the volcano continues to be somewhat active. The Montserrat government has tried to deal with this crisis by building a new city and port on the North West coast of the island away from the volcano.

This puts most claims of crisis into perspective: not only has the volcano devastated the housing stock, but it has also depleted the population and destroyed much of the infrastructure of the island.

- Imagine how a large country would cope with a crisis on a proportionate scale.
- How do the housing problems in your city or country compare with Montserrat?

Discussion points

1 Is there any problem in overstating your case if it means you win more resources for housing?
2 If we haven't got a housing crisis, then what have we got?

See also

Chapter 22 Sources of finance
Chapter 27 Boom and bust

Further reading

Oliver, P. (2006): *Built to Meet Needs: Cultural Issues in Vernacular Architecture*, London, Routledge.
Turner, J. F. C. (1976): *Housing by People: Towards Autonomy in Building Environments*, London, Marion Boyars.

Part 5
Money

22 Sources of finance

Housing policies, like our aspirations, are nothing more than fine words and pleasant dreams unless we are able to put them into action, and to do that we need money. Finance is what allows for the production and consumption of housing. We need money to build and maintain the nation's housing stock, but also to pay for it, in the form of rents, mortgages, loans and repayments.

There is a tendency to think that housing finance is all about government subsidies, such as capital grants, housing allowances and tax relief. These are indeed important now or were in the past. However, we need to be aware that there is more to housing finance than subsidies. The majority of households in the UK, US and parts of Europe are owner-occupiers who pay for their housing from their own income. Therefore much of housing finance is found privately, mainly from earned income. Of course, a household's income is normally used to repay a loan provided from a commercial bank or building society. This is another important source of housing finance. In addition, households use their own money and borrow in order to fund repairs and improvements to their dwellings. They also spend money on decoration, furnishings and fittings.

But private lending has also become increasingly important in rented housing. Both private and social landlords have to borrow from banks and building societies, just like private households. So we need to be aware that housing finance consists of more than subsidies from government. It involves the far larger sums spent by households and housing organisations that are derived from income and from borrowing.

But there are two further facets of housing finance that we need to consider. First, housing is a store of wealth, and thus we need to be aware of the fact that the housing stock is an asset that can be used by its owners. Individuals can, and do, tap into this wealth in order, say, to set up a small business, pay school and university fees for their children or enjoy their retirement. Landlords can use their assets as security for future development. Thus housing wealth can allow households and landlords to develop further housing and non-housing activities.

The second issue returns us to the role of government. Because housing is so expensive and so valuable an asset – as well as being so important to our well-being – government feels the need to regulate housing finance. It can do this through interest rates that affect mortgage repayments, by controlling rents through rent controls and regulating standards that impose costs on landlords. Therefore we need to consider not just the money that government spends on housing, but the costs that its actions impose on the various players involved in the production and consumption of housing.

So finance comes from a range of sources – private income, borrowing, released equity and government – and this can be affected by government policy itself. In addition, as much of housing finance derives from earned income and borrowing, the state of the economy, nationally and globally, is important. If people are losing their jobs, or their incomes are static, this will impact on housing. Likewise, if government reduces public expenditure, this can affect the level of subsidies offered to landlords and households. Housing is only one aspect of government policy and only part of the economy. It therefore has to compete with other areas of public policy for resources. Thus the level of political support that housing has, and more particularly the popularity of the respective housing tenures, is important in determining the amount of finance on offer to housing. Housing is very capital intensive, and so housing development often bears the brunt of government cutbacks. But also, where owner occupation is the dominant tenure, it is something simply too important for government to ignore, and so in difficult economic times it might actually receive additional support, even as subsidies to social housing are being reduced.

This tells us that government intervenes in different ways. It may have completely different strategies for the different tenures, offering support for one while cutting back on its commitments to another. But government's role also differs according to a household's income and therefore their ability to provide housing for themselves. In some cases government offers financial support and regulation, whilst in others (and this is the majority) it merely regulates standards. Moreover, this regulation might directly or indirectly impose costs on households, rather than providing them with financial support. Government intervention, and the level of financial and material support offered, will also differ according to the nature of the household, for example, families with children may be treated more favourably than single people.

So government intervention does not always reduce the costs of households or landlords. But, having said this, one of the main aims of government intervention is to make housing more affordable, and this is the purpose of subsidies. In simple terms, subsidies are intended to make housing cheaper and more affordable than it otherwise would be. Subsidies, therefore, are about altering the cost of housing and so potentially allowing more households access to it.

A rather more detailed definition is offered by Oxley and Smith (1996), who see a housing subsidy as 'An explicit or implicit flow of funds initiated by government activity which reduces the relative cost of housing production or consumption below what it otherwise would have been' (pp. 40–41).

This is a useful definition for a number of reasons. First, it is neutral, showing that subsidies can be used for all housing tenures. Whilst there is a tendency to concentrate on social housing, we need to be aware that governments also subsidise the private sector through housing allowances and improvement grants, and owner-occupiers through a variety of forms of tax relief and exemptions. Second, this definition does not just refer to the use of public funds. The reference to an implicit flow of funds can be seen as a reference to measures such as rent control, where private landlords effectively subsidise their tenants because they are not permitted to increase their rents above a ceiling set by government. What this shows is that a particular policy can benefit one group while adversely affecting another. The definition covers subsidies ranging from tax relief for owner-occupiers to government grants to housing associations. Finally, it demonstrates that subsidies can be directed towards landlords to assist them in building, managing and maintaining dwellings, but also to the consumers of housing in the form of tax relief or housing allowances. Subsidies can be used to support both production (supply) and consumption (demand).

Accordingly, any definition of a subsidy is also very much tied up with how it is used. For instance, subsidies paid to housing organisations, which allow them to build new dwellings at subsidised rents and to maintain their existing stock, have a markedly different purpose and effect on housing systems than subsidies paid to individuals that assist them in affording market rents. Subsidies to housing organisations assume that we need more housing and are therefore explicitly aimed at increasing the supply. Subsidies paid to individuals will not necessarily encourage an increase in the supply of housing, but are rather intended to help households afford what already exists.

Finally, we should note that some government intervention is driven by clear principles based around need and social justice. Government intervenes to help those in poverty, who are vulnerable or in severe need. But it might also be the case that its political objectives relate not to poverty reduction but to rather more general aims. In particular, politicians tend to describe owner occupation as an aspirational tenure and so will support it even if it means helping households who already have a sufficient income to fully fund their housing. We might suggest that this is not a terribly efficient or wise way to spend limited resources. But politicians also know that they can only achieve anything if they garner enough votes to bring them office. Accordingly, they will use all the policy and financial mechanisms available to them to ensure that they win and stay in office.

Discussion points

1 How important is finance to housing systems?
2 Does it matter where the money comes from and how it is paid?
3 What are subsidies for?

See also

Chapter 11 Owner occupation
Chapter 14 Social housing

Further reading

King, P. (2009): *Understanding Housing Finance: Meeting Needs and Making Choices*, 2nd edn, London, Routledge.
Oxley, M. (2004): *Economics, Planning and Housing*, Basingstoke, Palgrave.
Oxley, M. and Smith, J. (1996): *Housing Policy and Rented Housing in Europe*, London, Spon.

23 Markets

Most households, historically and geographically, obtain their housing through a market. Markets are therefore fundamental to housing. This does not mean that they always work well or that some people might earnestly wish for an alternative which they believe would work better. But markets are what we have and so we ought to try to understand something about them.

A market can be seen as a physical entity, a place like a street market where buyers and sellers come together within a defined space to buy and sell goods. This is the simple ideal of a market: where there are a number of traders competing against each other and so allowing customers to come to an informed decision on price and quantity. The customers are capable of making comparisons between each trader in terms of the price and quality of goods. But, of course, each trader is equally capable of seeing what their competitors are offering and can change their behaviour accordingly.

This ideal of a market supposes that certain conditions exist. First, there is the assumption that both consumers and suppliers have perfect information about the market. They either have, or can easily obtain, all the information on price and quantity in the market. Second, there are many suppliers so that no one person or group has market power and is therefore able to control the price. Any market trader would have to alter their prices to take account of their competitors, or at least they have to if they wish to sell anything because it is all too easy for customers to spend a few minutes walking around the entire market gaining all the information they need with regard to what is available and at what price. Third, it is assumed that the various players act rationally. It is held that suppliers seek to maximise their profits and consumers to maximise their utility, or the amount of benefit for any given amount of money spent. This means that all parties alter their behaviour according to market conditions: suppliers will supply more when prices rise and less when they fall; consumers will demand more as prices fall and less as they rise. Finally, there are no barriers to entry to the market. This means that it is straightforward for any person to set up in business, and no one who has the money is prevented from buying the goods, should they wish.

But, whilst it is convenient to use a place like a street market as the perfect example, it is by no means the case that all markets are like this. This means that some of the assumptions discussed above might not always hold good. In particular, it is now quite rare for a market to be contiguous and exist in one defined place. Most markets are not within a physically confined space where we can compare one supplier with another.

For example, we might now buy our apples from a supermarket instead of a street market, and we certainly do not drive from one superstore to the next to compare prices. Instead we might well stay with the one supermarket that we know offers the range and quality of produce we want. Yet, if it fails to do this and its prices no longer stand comparison with its competitors (which we know about because of advertising), we will change to another supermarket. Indeed, even in a street market there will be many regular customers who would walk straight past other stalls to buy their fruit and vegetables from their regular stall. But they only do so because they are certain that they will get what they want and, if they do not, then they could readily go elsewhere.

But it is no longer even necessary to leave the house to participate in a market. The development of Internet shopping and sites like eBay mean that markets need not be physical spaces at all. We can purchase our groceries on-line and have them delivered, and we can compare the price of the latest blockbuster novel or DVD on the various on-line sites and make a choice accordingly.

So, instead of seeing markets as physical spaces, they are best seen as a set of relations between people and companies and other organisations. Indeed markets are an example of human interaction. It is merely a convenient way of describing a set of transactions where money and goods change hands. It is where buyers and sellers come together to meet their needs to their best advantage. Hence, eBay is just as good an example as a street market.

We tend to talk about 'the market', but this is usually where we wish to make a political point, or to contrast it with government. In this way, the term is often taken as either a criticism or as a totem. 'The market' is either what impoverishes the developing world and is organised for the benefit of the filthy rich, or it is seen as an ideal form of social organisation and means of liberating individuals from the dead hand of government intervention. But, in either case, there is a tendency to preface the word market with the definite article 'the' as if it is just one, albeit very large, thing.

But it is not really proper to talk in this manner. A market can refer to all transactions – hence national and even global markets – or the term can be used for a specific subset of all transactions, such as the labour market or the housing market. But these subsets can and should be divided further. So we can talk of regional or local housing markets, or particular types of housing market, such as private renting in Leicester or owner occupation in Melbourne. Even within these sub-divisions we might be able to break them down further, as households in Leicester want particular types of property in specific parts of the city. So we can talk of the markets for three-bedroom houses and for one-bedroom apartments, which effectively have different customers and which may be affected by different factors, such that prices in one offer no real guide to prices in the other.

So, in terms of a definition of a market, we should not see it necessarily as a physical place, and nor is it just one thing. Taking these issues into account, we can concur with John O'Neill (1998), who states that a market economy can be defined as:

> those social and institutional arrangements through which goods are regularly produced for, distributed by and subject to contractual forms of exchange in which money and property rights over goods are transferred between agents.

> (p. 4)

This definition does not depend on any particular place or sense of a contiguous entity. Rather, it can enclose a global economy where goods and services are traded internationally, as well as the local market for one-bedroom apartments in Leicester. There are a number of elements of O'Neill's definition which we need to stress. First, a market is a social entity. It is a set of relations between households and businesses where different needs and desires are matched. Second, O'Neill stresses the institutional nature of markets. This does not refer to a physical place, but rather relates to the crucial point that markets need formal social, legal and political arrangements to underpin them. Markets only operate when there is a means of enforcing rights and contracts on each party to a transaction. Buyers and sellers need to be able to trust each other and to have a degree of certainty that the other party will deliver what they have committed to. This shows that the political use of the term 'market' that places it in opposition to government intervention is somewhat simplistic. Markets need a legal and political framework in which to operate, and this would apply even in the most liberal of market-based societies.

Markets therefore are a social arrangement that exists to allocate goods and services and the rights over them. The reason that such an arrangement is necessary is because there are never enough goods and services to go around and thus some form of mechanism is needed to allocate them. The crucial point, therefore, is that of scarcity: if there is a limited number of goods and services, which may have competing uses, there needs to be some reliable means for determining their most effective use.

Of course, this does not mean that allocations have to be made by markets, but merely that this is one of the main means by which they can be, and have been, done. Allocations can be made by central planning or by a dictator, but decisions will still need to be taken. The argument of proponents of markets is that they are the most efficient form of social arrangement for the allocation of scarce goods and services.

On the basis of this discussion, we can point to a number of functions that markets perform. First, they are a means of allocating scarce resources. If resources will always be relatively scarce, some means of distributing them is needed. In markets this is done through the price mechanism. Second, a market is a means of matching wants and needs at the lowest price and through the most efficient means. This is because markets encourage competition between suppliers, who are forced to reduce their prices, and thence their costs, to win a share of customers. Third, markets allow individual customers to make choices. An individual customer is not forced to purchase anything, but rather they are able to choose between a range of goods, depending on how much money they have available and what their priorities are.

We can see markets, therefore, as being about information-gathering and decision-making: they allow us to gather the necessary information to make an informed decision about meeting our needs and desires. Indeed, contrary to the ideal of perfect competition, the actual information required to make a market decision is rather simple: *How much is it?* As customers, we do not need to know how apples are grown or how they reach the market stall. All we need to be aware of is the price of apples and whether we can afford them. What this points to is the importance of price as the key signal of information in a market.

The importance of prices is that they provide the organising mechanism within markets. Price is essentially a means of signalling information to individuals and businesses about what is available and whether it is attainable. The key here is the relationship between price and scarcity. The price of something tells us how scarce it is and this acts as an allocation mechanism. Some may consider this to be unreasonable. They see the price of a house, realise that they cannot afford it and conclude that this is unfair. However, the problem is not the price, but rather the relative scarcity of housing of the type and in the location sought. Importantly, if price was not used to allocate the house, then some other means would have to be used, and this would be as prey to criticisms of unfairness as price. In any system there needs to be some means of allocating scarce resources. This does not mean that price is always the best means of doing so: it merely emphasises that, if it is not price, it will have to be something else such as bureaucratic decisions, political favouritism or a lottery.

As I have stated, there is a tendency to see markets in rather black and white terms: one either supports them or is opposed to them. The same applies, of course, to the role of the state. The reality, of course, is that any means of allocating resources is likely to be flawed. Government does not always achieve its aims, and politics can end in failure. Likewise, it is hard, in the early part of the twenty-first century, to argue for the virtues of housing and financial markets. The collapse of the sub-prime housing market in the US created serious repercussions in financial markets across the world. Many of the world's largest banks, like Credit Suisse and HSBC, announced huge losses on their investments and had to raise extra capital. Banks stopped lending to each other, and this led to the banking crises in Ireland, the UK and Spain. Governments had to intervene to prop up failing institutions and in some cases take them into public ownership. Whilst all this was happening, mortgage lenders increased their interest rates on borrowing and withdrew some of their products. Central banks reduced interest rates, yet this had no effect on the mortgage market, with lenders finding it hard to obtain new funds and increasing their rates to borrowers. Accordingly, in many countries house prices started to fall and many house builders reduced or even stopped new development, and the levels of activity in the housing market dramatically declined.

So, from this position, we could quite properly argue that housing markets do not work terribly well: many households, particularly the young, could not afford to become owner-occupiers and others were struggling to afford their mortgage repayments. Markets were failing to operate in the manner we might have expected and, this being so, we might argue that markets should not be the main means by which housing needs are fulfilled. What it certainly does show is that markets in reality are much more complex than the model of perfect competition would suggest.

Indeed the idea of *market failure* is commonly used with regard to housing markets. This is an important concept, in that it forms the major justification for government intervention: markets fail and so government has to intervene to regulate current provision or to increase supply. Market failure can be defined as where the conditions necessary for a market-efficient allocation do not exist. Market failure is therefore precisely when markets do not operate according to the simple model of perfect competition. It is when a market does not provide what is demanded at the cheapest possible price. This does not mean, of course, that markets do not work at all, but rather that they fail to be as efficient as economic theory states they could be. There are a number of causes of market failure and it is important to understand what these are, as they provide the main justifications for state intervention.

First, we can suggest that markets create *externalities*. This is where the actions of consumers and producers impact, positively or negatively, on a third party or on society more generally. We might see this as a social cost or social benefit that is borne by society as a result of private decisions taken by individuals and businesses. The classic example of an externality is industrial pollution caused by the production of goods and services. These goods are in demand and so suppliers are prepared to produce them. Yet the effect of industrial production is pollution, which places a cost on society as a whole. The cost of dealing with industrial pollution is a negative externality that has to be borne either by society, other individuals or businesses in other areas. In terms of housing, if we fail to maintain our property it can impact on our neighbours, have a negative effect on house values and quality of life, cause a nuisance to others and so on.

However, the problem with dealing with externalities, be they positive or negative, is that they are often difficult to quantify, as they frequently depend on the subjective perception of individuals. For example, my neighbour's attitude towards noise and nuisance may be very tolerant and so she is prepared to put up with my musical taste, whereas someone else might complain very quickly or even move away. What this suggests is that we can point to external costs, but perhaps find it hard to quantify them, and so struggle to deal with them in any systematic manner.

One means of dealing with externalities is to internalise them. This effectively means that, through regulation, the causers of an externality bear the cost. So, for example, we can insist that cars are fitted with anti-pollution devices, that hostels follow fire regulations, that environmental health laws are complied with and so on. The costs of dealing with an externality are now borne by the perpetrator. However, this does not deal with the problem of quantification, and regulation may go too far and impose a disproportionate cost that falls on those on low income who are less able to deal with changes in costs. An example might be the imposition of regulations on landlords which increase their costs to the extent that they leave the market and so reduce the availability of housing for low-income households. Crucially, there is no exact method for determining the proportionality of costs, in that landlords and their potential tenants have different thresholds regarding their preparedness to absorb additional costs.

An issue allied to that of externalities is that individuals acting in their own self-interest do not necessarily think of future generations. However, housing is a long-lived asset, which is expensive to provide and to maintain. Once housing has been provided, it is not easy to remedy its failings and it is wasteful to replace it. Most households live in 'second-hand' housing and many own dwellings that are older than they are. This means that the quality that we build housing to impacts not just on ourselves, but also on future generations. In general, the number of amenities in a dwelling increases over time, and so the need for space increases. But it is very hard to predict what future needs will be. Therefore a household might only be concerned with their immediate needs and not consider those who are not yet born who will expect amenities not yet invented. Yet, if a dwelling has to survive and be viable for a century, then some means have to be found to deal with the needs of future generations when we build today.

The cost of housing is such that it normally has to be financed by borrowing. It is very rare that we are able to pay cash for a dwelling; rather, we need to take out a mortgage with arrangements to pay off the loan over, say, 25 or 30 years. But this introduces uncertainty with regard to our future income, fluctuations in repayments due to changes in interest rates, changes in house values and so on. It is important to remember that when we buy a house we will not know exactly how much it has cost until we have made our last mortgage payment. Changes in interest rates will mean that our monthly costs might increase or decrease, and potentially the changes might be quite dramatic.

Even the most keen exponents of markets admit that they are not fair, in that they do not distribute resources according to merit or the amount of effort exerted or, indeed, according to need. Markets are impersonal entities to the extent that they guarantee no one anything just because they might need it. The support for markets is often based on the utilitarian argument that they lead to far better outcomes than any other form of social organisation. Markets are not perfect; they are just the least worst option.

Yet it might be that a society places fairness or social justice above economic efficiency or the entitlement to property. This society might find it unacceptable that some people, who were born lucky, have more than enough, whilst others, who have worked hard all their lives, end up with very little.

Some people inherit considerable housing wealth from their parents and this can be seen simply as a matter of luck. The inheritors have done nothing to deserve this wealth. A society might therefore seek to moderate these effects by intervening in a market.

This problem is exacerbated by the relatively high cost of housing, which means that there is a trade-off between quality and affordability, which raises issues of the distribution of resources. If some people are left poor as a result of the operation of a market, then a government might wish to control that market so that it works more for the benefit of the poor. Government, of course, might not succeed in its efforts, but this does show that there are other priorities than market efficiency.

The supply of housing in the short run is inelastic. This means that supply does not respond proportionately or immediately to changes in demand. It is not that flexible to changes, as it takes time to build new dwellings. Therefore, if demand for housing in an area increases, it does not follow that there is an immediate increase in supply. Indeed, increases in supply can take a number of years to come through. The result of this inelasticity of supply is an increase in price. There are a number of reasons why this inelasticity occurs.

First, there are limits as to how far we can increase the productivity of housing. The construction industry tends to be labour intensive, with the need for specialist trades such as bricklayers, plasterers and electricians. On top of this, the fallout from the failure of high-rise development in the UK, USA and Europe in the 1950s and 1960s led to a justifiably cautious approach towards experimental design. Planners and developers are rightly cautious of untried techniques and tend to trust what is known and, importantly, what is popular. Most households like traditional housing designs, such as low-rise housing located within green space. This might, however, make development more expensive, particularly the cost of land, and so extend the period in which housing supply catches up with demand.

Second, there is one crucial way in which housing differs from most other goods and services. Housing lacks mobility, in that it cannot be moved from one place to another in response to changes in demand. If there is a shortage of apples in one city, a new supply can be transported in by plane and truck relatively quickly. However, a shortage of housing can only be alleviated by building more housing in that area. The alternative is to encourage or persuade some households to move to other areas where housing is in plentiful supply.

Third, there is what might be referred to as spatial and situational restrictions on the supply of housing. This includes planning restrictions on development which seek to protect other households. Households can only extend their dwelling or developers only build a new estate if it is reasonable and does not impinge on others. A relevant issue is that land is scarce, particularly in urban areas, and so there might be competing uses. Large companies, such as supermarket chains, want to be located near housing and transport systems, but local residents might object to this development. Other residents might treasure the green space that surrounds their housing and therefore complain at proposals to develop on this space, even though there is unmet demand for housing. Clearly, land availability is an important restriction on housing development. Government might seek to deal with this by encouraging building on brownfield sites by offering subsidies or setting targets.

Some parts of the country will be seen as more desirable locations than others, with the result that housing is more expensive in these areas. In some locations there might be a relatively high proportion of the housing that is used for holiday lets or second homes. These dwellings might be unused for much of the year, as well as having the effect of increasing overall house prices in the area. This is particularly a problem in rural areas, where average wages are low, or in seaside areas where employment is seasonal. The effect of second homes might therefore be to price local people out of the housing market. Indeed, the market signals offered are precisely for more holiday accommodation rather than housing for low-income households. All these factors can limit the supply of new housing or cause delays in increasing supply. As a result, government might feel it needs to intervene to speed up the process, either by directly subsidising provision or by easing planning restrictions.

Of course it is all too easy, especially after a recession, to criticise markets. However, we have to recognise that, after 2008 most markets carried on working. They were not perfect and they did not satisfy everyone. However, households could still buy, sell and rent housing and this has become easier as the economic conditions improved. It is true that these conditions were improved by government intervention and that there are still problems which may take considerable time to sort. But this returns us to the politics of housing markets. Both the proponents and opponents of markets see markets in terms of the simplistic perfect competition model. Markets are taken to be the opposite of government and to work in a fundamental way. Yet markets do not work in this manner, they probably never really have and are unlikely to do so in the future. The reality is that markets operate in conjunction with government. Consumers need to trust that they will get what they have paid for and need redress when they do not. Contracts need to be enforced and rights protected. This is, and always has been, the role of the state.

The question therefore is not really whether we want markets *or* the state, but how much of each: what is the balance that we think is ideal to create the type of society that we wish to live in? This differs from country to country, based on history and culture, but it will always be some combination of market and state.

Think piece: Knowledge, planning and prices

A market economy depends on what Thomas Sowell terms systemic causation: this involves 'complex reciprocal interactions' (2007, p. 63). This is when the behaviour of one element alters another and is in turn altered by it. There is reciprocity in the interaction, such that all elements behave differently; there is no straightforward causality. This means that we have to be concerned with what emerges, rather than what was intended. It also means that the impact of any one decision is small and ineffectual. It is the combination of all these decisions that has the real impact. This means that decisions cannot be taken on a whim or even by a sustained act of will. While all decisions are consequential to the individual that takes them, they might not necessarily appear so important to the market or to other consumers.

A key thinker in how markets actually work is the Austrian economist and social philosopher, Friedrich Hayek (1948, 1967, 1978, 1988). He is concerned with the nature of decision-making in a complex modern society. Hayek makes the distinction between two types of rationalism, which he calls constructivist and evolutionary. Constructivist rationalists believe that all human institutions and behaviour are the result of human reason and human will. Human beings can therefore master society and come to control and reform it. It is the belief that human institutions can be constructed to achieve particular desired aims. This is the mentality of the central planner trying to set all the prices in an economy. Hayek contrasts this with evolutionary rationalism, which is an attempt to understand how civilisation has developed, but with the attendant recognition of the limits of our current knowledge and thus the likelihood of unintended and unforeseen consequences. According to Hayek, evolutionary rationalism assumes that knowledge accumulates through individual action and experience, but in an unpredictable way. This implies that social change is better seen as a slow development caused by an untold number of actions, and not as the result of deliberate acts of manipulation.

Accordingly, Hayek suggests that social evolution, including the development of markets, occurs through human action, but not by human design. Here he is concerned to correct what he considers to be the fallacy of constructivist rationalism which, he suggests, pervades collectivist approaches to social and public policy. This is the belief that, as all institutions have been made by human action, they can be remade by further deliberate action. This, according to Hayek, is a fundamental error on the part of policymakers. Institutions have developed in a particular way out of the uncoordinated interrelationship of millions of individual actions, none of which in themselves were of fundamental significance. This is a similar point to that made by Sowell when he states that no one individual is capable of altering the prices, yet markets are merely the combination of millions of individual decisions in reaction to price.

Hayek's argument is based on one of the foundational concepts of market analysis: the metaphor of the invisible hand. This term was first explicitly formulated by the Scottish economist and moral philosopher Adam Smith in the eighteenth century (1976). He used the term to suggest a socially beneficent outcome from self-interested actions. He saw that markets were not centrally co-ordinated but were rather the accumulation of individual actions by individual consumers and producers, each seeking to maximise their outcomes. Beneficial outcomes were arranged *as if by an invisible hand*. We should note that Smith was not suggesting that the invisible hand actually existed: it is metaphor for how a market works for the benefit of a society, without any central direction and without any one individual intending to meet any needs but his or her own.

This idea of the invisible hand provides a very sophisticated model for how an economy operates. In particular, it allows us to see that market outcomes are essentially the result of unintended consequences and the crossover effects of decision-makers operating with only limited knowledge, namely the price of things. It tells us that markets have not developed as a result of specific policies or government action, but rather as the result of countless individual decisions over centuries. Each of these decisions was of little consequence, but each has some impact in determining the subsequent actions of others. It also shows that markets are hugely complex. Yet this complexity is itself based on the simple notion of prices determining individual actions within a market.

References

Hayek, F. (1948): *Individualism and Economic Order*, Chicago, IL, Chicago University Press.

Hayek, F. (1967): *Studies in Philosophy, Politics and Economics*, London, Routledge.

Hayek, F. (1978): *New Studies in Philosophy, Politics, Economics and the History of Ideas*, London, Routledge.

Hayek, F. (1988): *The Fatal Conceit: The Errors of Socialism*, London, Routledge.

Smith, A. (1976): *An Inquiry into the Nature and Causes of the Wealth of Nations*, 2 vols, Indianapolis, IN, Liberty Fund.

Sowell, T. (2007): *Basic Economics: A Common Sense Guide to the Economy*, 3rd edn, New York, Basic Books.

Discussion points

1   How realistic is the model of perfect competition? Why do we still use it?
2   Is it unfair that we cannot afford a mansion, when some people can?
3   In what ways other than markets might we allocate scarce resources?
4   If markets fail, why do we still have them?

See also

Chapter 8 Choice
Chapter 11 Owner occupation

Further reading

King, P. and Oxley, M. (2000): *Housing: Who Decides?*, Basingstoke, Palgrave Macmillan.
O'Neill, J. (1998): *The Market: Ethics, Knowledge and Politics*, London, Routledge.
Oxley, M. (2004): *Economics, Planning and Housing*, Basingstoke, Palgrave.

24 Rent

One of the arguments in favour of owner occupation is that, having paid your mortgage for 25 years, you now have something to show for it. You own the house and it is yours to do with as you will. However, if you have been renting for 25 years what, so the saying goes, have you got to show for it? Renting is, to follow this line of arguing, throwing money away. Instead of investing in property, you have given the money away to some unscrupulous landlord.

This is a common argument, but it is also a strange one. The same people who argue that rent is money thrown away are unlikely to apply the same argument to food that we have purchased over the last 25 years, or the clothes we have bought in that period. Yet, where is the food now (don't answer that!). There are many things that we use and which we enjoy and, more importantly, that we need, which are no longer around. The evidence for all that food and that clothing is that we are still alive and healthy. And the same applies to the rent we have paid. Paying rent to our landlord has allowed us to live in secure accommodation, to raise our children and to flourish. We do actually have a bit to show for it.

So what then is rent? Quite simply it is what we have to pay a landlord to live in their property. It can therefore be described as a user charge that allows us – the user – exclusive rights over the dwelling for the period of the tenancy. In this way it is the same as buying a cinema ticket that allows us to use a particular seat in front of the screen or a train ticket that allows us a seat for the duration of a journey. Rent is actually an example of a very common transaction where the owner of a good or service allows us access in return for an agreed payment.

What often colours the discussion on renting is the dominance of owner occupation. This is taken as the normal tenure, and the others are judged accordingly. Yet, for much of human history, renting was more common than owning and, in some parts of the world, including affluent European countries, it is still as common to rent as it is to buy.

Rent can include a number of different elements. In some circumstances it might simply be a charge for the use of the dwelling. However, it could also include an element for the use of services, such as common facilities, cleaning, gardening, etc. It might also include utility bills such as fuel, telephone and Internet access.

One of the key political issues about renting is that rents tend to change more readily than supply. If there is an increase in demand for rented accommodation, it is often difficult to create additional supply. This is because of planning constraints, the availability of land and the length of time it takes to get on site and build. The result is that rents have a tendency to rise.

This situation can be affected by the state of other markets. For example, if there is a shortage of social housing, low-income households will have to compete for private rented housing. Likewise, in areas of high house prices, some groups may be unable to buy and so have no recourse but to rent. This can be a particular problem for first-time buyers who need to put down a sizeable deposit before they can gain a mortgage. These constraints on other tenures mean that there is not always a great deal of mobility between the tenures, and this means that rents in the private sector do not always reflect what is being charged in the other tenures. The result is that private rents can be much higher than in the social sector and also higher than monthly mortgage payments for a similar property.

These problems have led to frequent calls for rents to be controlled, either by statute, with the state imposing rent ceilings on private landlords, or by administrative means such as limiting eligibility or access to housing allowances. Rent controls can be an effective way of lowering the real term costs of renting, and they have been used extensively in many countries across the world. However, they do come with a cost in that they reduce the incentives for landlords to stay in the market as they cannot make a decent return on their investments (see Chapter 15, Private renting for a fuller discussion).

Much of the discussion has been on private renting, but we need to acknowledge that social housing is rented housing too. There is, though, a fundamental difference, namely that social housing is so organised to prevent landlords making a profit from renting. This means that rents are set to meet reasonable costs, which may include an element for future investment and for the repayment of debts, but this does not include any return to shareholders.

This raises a fundamental issue with regard to renting: is it ever justifiable for a landlord to make a profit out of the housing needs of others? We might argue that housing is a basic need and that it is immoral for anyone to make money out of vulnerable households who have no alternative access to housing. Hence it has not been uncommon over the last 150 years to hear calls that private landlords exploit working-class and poor households and that this should not be accepted in a civilised society. These calls for change have often increased when rents are rising due to shortages in supply.

However, we need to return to the start of our discussion and recall that there are many other goods and services that are as essential to us as housing. We need food and clothing in order to stay alive and to flourish, yet we obtain these through a market. Some households receive income maintenance payments to help them, but there is no call to take the whole food and clothing supply chains into public ownership. Indeed, the only manner in which we could insist on non-profit-based rented housing would be for all privately rented housing to be nationalised. This may be seen as desirable for some, but it is not a politically acceptable option for most countries (if only because there would be a legitimate fear that the nationalisation would soon extend to owner-occupied housing as well: why allow developers and sellers to make a profit and so 'exploit' buyers?).

However, in mitigation we can quite properly argue that markets for private rented housing are not the same as those for food and clothing. There tends to be much less opportunity for competition, and the supply constraints are rather different. As a result, we might argue that private renting needs rather more regulation than some other markets to ensure that the charges to the users remain fair and affordable.

Discussion points

1   How does renting differ from owning?
2   Is housing different from other goods and services and should it be treated differently than these goods?
3   Is private renting exploitative?

See also

Chapter 14 Social housing
Chapter 15 Private renting
Chapter 23 Markets
Chapter 26 Affordability

Further reading

Albon, R. and Stafford, D. (1987): *Rent Control*, London, Croom Helm.
Dorling, D. (2014): *All That is Solid: How the Great British Housing Disaster Defines Our Times, and What We Can Do About It*, London, Allen Lane.
Sowell, T. (2007): *Basic Economics: A Common Sense Guide to the Economy*, 3rd edn, New York, Basic Books.

25 Housing allowances

Since the 1970s there has been a shift in most developed countries away from object subsidies and towards subject subsidies. Instead of providing subsidies to allow landlords to build new social housing, governments shifted to supporting individual households on the basis that increasing their income would allow them to purchase housing in a market.

Peter Kemp (1997) has suggested three reasons for this general shift away from object subsidies. First, he cites the end of massive housing shortages in the 1970s as a result of the mass building programmes undertaken in Europe and elsewhere since 1945. By 1980 there were many countries that had a crude surplus of dwellings and this allowed governments to argue for a shift in priorities. Instead of the problem of shortage, governments' attention now turned to the shortage of income of some households as the key issue. Hence, there began a shift towards income maintenance and the use of housing allowances as a means of ensuring that low-income households could gain access to housing.

Second, Kemp pointed to the general economic malaise of the 1970s, with high levels of unemployment coupled with high inflation in many countries. The result was a belief that the welfare state was unaffordable in its current form. Already certain demographic trends were becoming evident, particularly longer life spans, and therefore it was felt that the welfare state was becoming an increasing burden at a time when governments were struggling to understand changes in the world economy. As housing, along with road building, is a very capital-intensive activity, it was an easy target for cutbacks.

Third, Kemp points to a change in the political climate, as well as in the economy. He suggests that there developed a general belief in market solutions to problems in social and public policy, emphasising the importance of the consumer over the producer of services. This was manifested by the election of right-wing governments in the USA, Germany and the UK, which survived for all of the 1980s and into the 1990s. This shift in political opinion was driven in part by the failing economy and the belief that the interventionist economic policies of the post-1945 era were no longer valid. But there was also an undoubted intellectual shift in favour of markets and smaller government.

This change in the balance between object and subject subsidies implies a change in the purpose of housing subsidies. Instead of subsidy being used to increase supply, it is now aimed at bolstering demand. The belief is that there is enough housing for the number of households in the country. What is therefore at issue is not the quantity of housing, but whether all households can gain access to housing of sufficient quality.

But the two different forms of subsidy are based on two different sets of assumptions about the role of government and the competence of individual households. To favour demand-side subsidies is to suggest that the problem is one of a lack of income and that individuals are basically capable of choosing if given the resources. Supply-side subsidies carry the implication of a more fundamental problem that cannot be solved by increasing household income alone. Individual households are not capable of influencing general social forces such as poverty and inequality, and so government needs to step in on their behalf.

There are a number of reasons given to justify the preference for subject subsidies. These are often discounted by academics and housing professionals, yet, as we have seen, these groups have essentially lost the argument, with the view of politicians favouring housing allowances dominating housing policy since the 1970s.

The first argument used to justify subject subsidies is the problem of producer capture. It is assumed that the purpose of subsidies is to help people in need. Yet subsidies are also used to control provision and to ensure that systems operate in particular ways. The question we therefore need to consider is who or what are subsidy systems for, and, in particular, do they benefit the producers of the service or the consumers? It can be argued that object subsidies can be controlled by producers and operate to their benefit. If producers can control subsidies – because they are paid to them – how can we ensure that consumers are being treated properly and that provision is being made efficiently and fairly?

One way of examining this issue is through the arguments of public choice theory. This is based upon three main criticisms of the role of public organisations. First, it is suggested that many public services are provided by monopoly suppliers, either at the national level, such as the NHS, or locally, such as local authority housing departments. Public monopoly can lead to poor performance because officials have little incentive to keep costs down or innovate. There are few financial or other benefits for those who innovate, and resources are not directed by the users but by a 'political' sponsor. Therefore officials are more likely to respond to political pressure than to that from customers.

Second, there is an absence of valid indicators of organisational performance by which to judge outcomes and ensure that consumers' interests are uppermost. Public choice theorists suggest that there are no unambiguous indicators in the public sector, such as profit and loss, making it difficult to evaluate individual or collective performance. Third, the large size of public organisations creates problems of co-ordination and control, and these lead to a decline in performance as the size of the organisation increases. In response to these issues, public choice theorists advocate a more competitive structure, with rivalries within the public sector and between public and private sectors. This would force greater information-sharing to enable performance to be judged and would break up large agencies into smaller units. Consumers could also be given some level of choice in determining their supplier and the level of service they receive. The most direct manner of achieving this is through the use of vouchers for services or by directing subsidies to the consumers themselves, thereby forcing producers to compete for their custom.

One of the main justifications for subject subsidies is that they can be targeted at those in need and then be withdrawn when income increases. Households who are allocated social housing can stay there all their lives, regardless of how their income and personal circumstances change. Thus, needy low-income households might be denied access to social housing because more affluent households remain in occupation, even though they might now be able to afford owner occupation or private renting. A system of subject subsidies, however, could prevent this because households are subsidised according to their current, and not their past, circumstances. The subsidy can thus be withdrawn if and when their circumstances change.

There has been some discussion in both the UK and the Netherlands about time-limited social tenancies and the potential of making them as means tested as housing allowances. This is still, however, a controversial proposal, with some people suggesting that it will merely worsen the problem of social polarisation in social housing, in that economic dependency will effectively become a condition for maintaining a tenancy. There might also be problems with moral hazard in that tenants would have an incentive to remain economically dependent in order to keep their social tenancy.

It could be argued that object subsidies give too dominant a role to landlords at the expense of tenants. Landlords are able to exercise control over rents and the level of service offered to tenants. However, paying subsidies to tenants gives them some negotiating strength in relation to rent levels. It would create a different and more equal relationship between landlord and tenant. In principle a system of housing vouchers, which operates in some parts of the US, allows for this situation. However, in practice it does depend on the co-operation of landlords and their preparedness to accept low-income tenants with vouchers.

A further advantage claimed for subject subsidies is that they can be tenure neutral, in that they can be applied to all housing sectors, including, if so desired, owner occupation. Subject subsidy systems can be devised that are so designed as to allow access to all and can be dependent only on income rather than tenure or any particular relationship with the state.

Whilst the supporters of object subsidies argue that they help landlords to build good quality housing, there is no automatic link between this form of subsidy and quality outputs, even where there is sufficient demand for the dwellings. Social landlords have been guilty of building poor quality and unpopular housing, with many social landlords encouraged by the subsidy system in the 1950s and 1960s to build high-rise blocks, which are not universally popular and, as with the example of Ronan Point in London that collapsed in 1968, have proved on occasions to be disastrous.

Developing this point on the nature of the outcomes of state provision, it can be argued that object subsidies have led to ghettoisation and unbalanced communities. They have created large estates where many of the occupants are economically inactive and where those who can afford to leave do so. As a result, social housing has become a key indicator of social exclusion.

But perhaps the most significant benefit to be derived from a subject subsidy system is that it can offer households some choice over where they live and the type of accommodation they wish to reside in. Paying the subsidy directly to households enables them to exercise more control over their lives than if the subsidy were paid to landlords who then build what they feel is required. Paying benefit directly to individual households enables them to have some choice over their housing that is not open to households in an object subsidy system. Of course, this does not mean that households have an untrammelled choice or that their options are limitless. This is sometimes used as a criticism of choice-based systems: because choice is not limitless, and indeed in practice might be quite restricted, it is somehow an illusion. Yet the choices open to all households, even to an extent the wealthy, are limited, being hemmed in by income and family ties, employment opportunities, available schools and the quality of public transport; this is before we even consider such issues as housing supply and availability. What we have to remember is that choice does not have to be limitless to still be choice.

In practice, the situation may not be as simple as suggested above. This is because many housing systems have to deal with a legacy of the past. In some countries there is a significant stock of social housing and there may still be outstanding debt attached to these dwellings, which need continued government support. In addition, it might well be that tenants in social housing, because they are on low income, are themselves in receipt of a housing allowance. This means that housing allowances are used to subsidise social housing rents that were set at below market levels because of another form of subsidy. In many countries the shift from object to subject subsidies is not total or complete and therefore the two forms of subsidy become linked and start to impact on each other.

However, this does not negate the shift towards subject subsidies as part of the general shift towards choice and considering tenants as consumers rather than passive recipients of services. This shift has been on going in many countries for more than a generation and its effects are now considerable.

Think piece: Just a matter of money?

Most people buy their own housing using their own income. They rent, buy and sell in a market on the basis of price, competing with other households for housing. It might therefore be argued that supplementing the income of poor households will allow them to compete with the majority. There is not, then, a housing problem but an income problem.

But if you were to ask most academics whether this is the case, most would argue that it is not. They would suggest that the problem of access to decent quality housing goes beyond having money in your pocket, but relates to more fundamental structural factors in a society such as inequality, race and class.

But if having an adequate income is sufficient for the majority, why is it not for those on low incomes? What is the difference here?

Discussion points

1   What are the main advantages of subject subsidies?
2   Has the move to subject subsidies created a fairer housing system?
3   Are tenants customers?

See also

Chapter 8 Choice
Chapter 14 Social housing
Chapter 15 Private renting
Chapter 24 Rent

Further reading

Kemp, P. (1997): *A Comparative Study of Housing Allowances*, London, HMSO.
Kemp, P. (Ed.) (2007): *Housing Allowances in Comparative Perspective*, Bristol, Policy Press.
King, P. (2006a): *Choice and the End of Social Housing*, London, Institute of Economic Affairs.

26 Affordability

Affordability is an important concept that is now commonly used by housing practitioners, commentators and academics. However, it is not a concept used in common speech. One would not usually say that one lived in 'affordable housing'. In essence the term has come to be used as an alternative for rather older and more established terms such as social or public housing. It is used in place of these terms because it is tenure neutral. It can refer to social and private renting as well as to owner occupation. Instead of referring to the ownership of the housing, it refers to the fact that it is being offered more cheaply than other forms of housing.

The fact that anyone other than housing professionals and commentators does not use the term, including those living in it, has meant that 'affordable housing', 'affordable homes' or simply 'affordable' has become almost a technical term. We hear commentators talking about the number of affordable homes built as a percentage of the total stock on an estate: we hear the phrase 'with X percent being affordable'. To those who are not *au fait* with housing jargon this can sound absurd. One might wonder why anyone would even contemplate building housing that was not affordable.

But, as a technical term, it has come to have a particular meaning. It is not that people cannot afford to live in housing, but rather the right sort of people cannot live in it. So a house can be defined as 'unaffordable' even though it is lived in and many people could comfortably afford it.

One might have thought that any dwelling is affordable if there is someone living in it in a reasonably sustainable, long-term way. If they can pay the rent or mortgage every month, pay their other bills, and do this over time, then surely that dwelling is affordable. Were one to argue this with a housing expert, however, one would be accused of being naïve and not understanding the problem.

The reason for this is that the technical term 'affordable housing' has another element to it that nearly always remains unsaid. What is really meant is whether the dwelling is affordable for those on low incomes. It has a quite specific meaning relating to whether households on the lowest incomes could afford to live there. If they cannot, then the housing is deemed to be unaffordable. Hence it is possible for housing experts to claim that, in a city such as London, where the demand for housing outstrips supply, and therefore the housing market is very active, the majority of dwellings are unaffordable. The housing can be completely occupied by people who can afford to pay the going rate, but it can still be deemed unaffordable.

We might actually offer an alternative definition of affordable housing, as that housing which receives a subsidy to make it affordable for those on low income. The problem of unaffordability for some households calls for subsidisation and regulation rather than relying on the market to allocate scarce resources. Hence we might also define affordable housing as that which is offered at below its proper market value.

However, all we have done here is to define the term. What we have not done is consider what it actually means in practice for housing to be affordable. The most straightforward way that affordability is defined is to express it as a percentage of earnings, so, if a household has to pay more than a given percentage of their earnings on housing, then that dwelling is unaffordable. Of course, it should be immediately apparent that what is being measured here is the income of the intended users and not the value of the house. It might be possible to come up with some general figures based on average earnings for particular needs groups and values for certain types of housing in a particular area. But, then, this is only relevant if that housing is then occupied by the correct needs group and, of course, if the household remains within the average income of that group. This is indeed a problem, in that the income of a household is determined at the application stage and might not be assessed again (although this depends on whether the tenancy is a fixed term: no such facility is available, of course, for housing for sale).

So housing is affordable if housing costs are below a certain proportion of earnings. But how do we decide on what is the correct percentage? We cannot do this on the basis of what is the market average, because this is precisely the problem: there are a number of households who cannot afford these levels. In any case, we have to realise that different groups in a society will either be prepared to or are capable of paying a higher proportion of their income on housing costs. Those on high incomes might be able to afford an unusually high percentage simply because this will still leave them with sufficient residual income. A billionaire could pay more than 70 per cent of their income on housing and still have plenty left for the yacht and the champagne parties. The same percentage would be disastrous for those on minimum wages.

It might also be the case that first-time buyers might willingly pay more than the affordable percentage. This is because they will tend to be younger and at the start of their working lives. They can therefore reasonably expect their incomes to rise and so be able to afford their housing costs in the long term. Also, households with double earners might be able to afford higher housing costs, and therefore we need to be aware not just of average local earnings but actual household incomes.

A further problem is to actually determine what to include in any calculation of housing costs. Should this include only direct mortgage and rent payments, or should we include service charges, utility bills, insurance and repair bills? Clearly how we define housing costs will impact on what is deemed affordable.

All of these are largely technical questions and it is certainly possible to come up with decisions on what we mean by housing costs and find some consensus on what would be a suitable percentage. What we have to conclude, however, is that this is by no means an exact science and there will be some controversy as to what is or is not affordable. So we need to be careful over how we use the concept and perhaps show some awareness about the ambiguity and, for some people, the absurdity of the concept.

A final point to make is that a considerable number of those living in affordable housing, which has been set up to ensure that rents do not exceed a given percentage of local earnings, will also be in receipt of a housing allowance to help them pay their rent. What this means is that, even this so-called affordable housing is in fact only affordable with an additional subsidy, which in some cases will actually cover the total cost of their rent. This means that any definition of affordability is entirely moot, and all our discussion has been for nothing.

Discussion points

1   Is a house really ever unaffordable?
2   Why have we stopped talking about social housing and replaced it with affordable housing?
3   What should we include in an affordable rent?

See also

Chapter 22 Sources of finance
Chapter 23 Markets
Chapter 24 Rent
Chapter 25 Housing allowances
Chapter 34 Planning

Further reading

King, P. (2012): 'The Ethics of Affordable Housing', in Chadwick, E. (Ed.): *Encyclopedia of Applied Ethics*, 2nd edn, San Diego, CA, Academic Press, vol. 1, pp. 72–78.

27 Boom and bust

The 2008 financial crisis was not the first and probably will not be the last time that a tumble in house prices has followed a steep rise in prices. Indeed, the history of housing markets is pretty much one of booms followed all too regularly by a bust. This is something that is not unique to one country but can be found throughout the world. Indeed, more recently boom and bust have been international phenomena, with the only difference that some countries have fared rather worse than others.

But why does this cycle of boom and bust keep happening? Why haven't we learnt to prevent them from occurring? One of the key reasons is precisely that politicians and bankers have often claimed that they do know how to prevent them and so act as if they won't happen again. Their confidence leads them to take risks and act injudiciously, with the result that it all comes crashing down. But it is not entirely due to human fallibility. One of the problems — and one of the reasons why people think they have found a cure — is that the exact cause of a bust differs on each occasion. This means that we cannot necessarily learn from the last one, except in the most general sense that we ought to be cautious and not boast too much about our economic competence.

A second problem is that boom and bust happen just seldom enough to mean that the people running the banks and governments now are not those who were around at the time of the last crash. Those running banks have probably all been sacked or pushed into retirement, and the politicians have lost office. They have been replaced by younger models who think they know better and swear they will never allow a crash to happen again. But, as a result, these people actually lack any experience of a financial crisis and so soon start making their own mistakes.

However, even though we cannot pre-empt the cause of the next boom and bust, we can point to a few general points that appear to be typical. First, as housing is funded by borrowing and the sums involved are so large, housing markets are linked into international financial markets. This means that housing can be affected by shifting patterns in the world economy and vice versa.

Second, it is in the nature of housing markets that supply lags behind demand. Demand can change much more readily than supply, and so when demand does increase the result is not an immediate increase in supply but a rapid rise in house prices. This can lead to a situation of what economists call perverse demand. In normal markets a rise in prices will led to a fall in demand. However, it is common for the opposite to occur in housing markets, such that, as prices rise, demand rises along with it. This may seem odd but the reason for it is quite simple: households can see prices rising and they expect them to continue increasing into the future. They reason therefore that housing will continue to become more expensive and so it is entirely logical to bring forward plans to buy or to move. As with the banks discussed above, busts happened infrequently enough for households to forget what happened in the past and to believe that the increase in house prices is a permanent phenomenon.

Third, markets, of whatever type, are essentially based on nothing more than confidence. A house has a particular value because someone is prepared to pay that price. Values are not intrinsic but are subjective, based on a particular individual's or group's expectations, aspirations and income. If no one believes that a house is worth as much as it was yesterday, then quite simply it isn't. These shifts in confidence can snowball when people start to expect prices to fall: the expectation inevitably proves correct, and this is precisely because people alter their behaviour to match what they now expect. So, if banks lose confidence in the market and each other, they will reduce their own borrowing and be less likely to lend to others. Thus, the whole financial system comes to a halt, leading to the weaker institutions actually failing. Busts can therefore be self-generating and they become hard to stop, as they feed on themselves in a downward spiral of falling prices and confidence.

This is a serious problem because one of the main functions of banks is to circulate money, without which nothing can be bought and sold. It has become a cliché that some banks are too big to fail, but the problem is really that they are too important to fail. It is not just that we have a mortgage with these institutions, but our employers pay our salaries directly into our bank accounts; we use their cash machines to get access to our money and we rely on banks to look after our savings. This means that, even if banks behave irresponsibly, it becomes necessary for governments to bail them out and support them.

Financial services is now a massive industry and in many countries it forms a significant part of the economy. Also, when the economy is working well, banks and financial services tend to be very profitable. This is at a time when more traditional industries such as manufacturing have declined. This means that finance provides an increasingly significant part of government tax revenues. Accordingly, governments have every reason to encourage financial services, even to the extent of lightening regulation and downplaying any risks.

Running alongside this increase in the importance of financial services has been a cultural change in many people's attitude toward debt and borrowing. While a few generations ago more emphasis would have been put on saving up for a major purchase, it is now more likely that we will take out a loan or use a credit card. Most developed societies, with the exception of China and Japan, now save considerably less, and prior to the 2008 crash the savings ratio (the proportion of our income that we save) was actually negative in some countries. Increasingly, we live in a culture where we expect to have access to goods and services now rather than waiting, and this applies to housing as much as anything else. Of course, this propensity to borrow is encouraged by the financial sector as it allows it to grow and become more profitable. Likewise, it tends to be supported by politicians who see a growing economy and optimism amongst the electorate and seek to profit from it politically.

Governments in many countries have also been fulsome in their support of housing markets and owner occupation in particular. There has been a long history of supportive rhetoric, whether it is in terms of creating a 'property-owning democracy' or building 'the American Dream'. But this has also been backed by financial support such as tax relief, direct subsidies to first-time buyers, government-backed insurance for lenders and more direct intervention such as the Right to Buy in the UK allowing social tenants to buy their dwelling. Owner occupation is seen as the aspirational tenure and this leads politicians to support it and seek to ensure that all those who want to become owners can do so. The result is that perhaps some marginal households enter the market who are not able to cope with changes in interest rates or periods of unemployment.

We can conclude, therefore, with the statement that booms and busts occur because no one has any incentive or good reason to stop them. In hindsight it may become clear that the logic was faulty and that households, politicians and banks were acting irrationally. But at the time they all thought they were being rather clever, and there was no one willing to challenge them.

Discussion points

1 Is boom and bust the result of not enough government regulation or too much?
2 Would it be possible to break the link between housing and international finance?
3 After all the problems caused by boom and bust, why do governments continue to support owner occupation?

See also

Further reading

King, P. (2010b): *Housing Boom and Bust: Owner Occupation, Government Regulation and the Credit Crunch*, London, Routledge.

Richardson, J. (Ed.) (2010): *From Recession to Renewal: The Impact of the Financial Crisis on Local Government and Public Services*, Bristol, Policy Press.

Shiller, R. (2008): *The Sub-Prime Solution: How Today's Global Financial Crisis Happened and What to Do about It*, Princeton, NJ, Princeton University Press.

Sowell, T. (2009): *The Housing Boom and Bust*, New York, Basic Books.

28 Borrowing

Housing is expensive to build, to buy and sometimes to maintain. This means that developers, landlords and households will need a stable form of long-term finance to help them achieve their ends. Fortunately the nature of housing as an economic good makes borrowing eminently possible. Housing is a long-lived asset, which can be expected to survive longer than even the longest mortgage or loan. Second, housing tends to appreciate in value, making it attractive as security for a loan. Indeed, this quality of housing is important in that it means that a dwelling can be remortgaged as its value increases, allowing the owner access to capital for other projects (or a nice holiday!).

So, borrowing is ubiquitous to housing systems and occurs at all levels. What differs, of course, are the sums borrowed by the various different players. It would be very rare for any household to be able to buy without recourse to borrowing and this would apply to those trading up and those buyers entering the market for the first time. The amount that households need to borrow obviously varies considerably dependent on the location of the property, its size and the percentage of the total cost that the household need to borrow. Mortgage lenders, even in developed countries, have traditionally been locally based and therefore quite small. In addition, they had to rely on savings from depositors as the major funds for loans. However, it is now more common for mortgage lenders to borrow from other financial institutions and then lend the money on to households. This is possible as they are able to borrow relatively larger sums at one rate of interest and then lend the money in small tranches to households at a higher rate. Through borrowing they are able to raise far more capital than they would from relying on savers and so expand their businesses. However, as a result they have become linked into global financial networks, which has proved disastrous for some lenders when the financial markets collapsed in 2008.

Both private and social landlords will also borrow money to build and buy housing. The scale here ranges from buy-to-let investors, who perhaps purchase just one or two dwellings as an investment, through to large commercial property firms who may develop and own hundreds or thousands of properties, perhaps in several countries. Social landlords will also borrow money to fund their development, even though they may still receive some direct (grants or cheap loans) or indirect (free or cheap land) subsidies from government. Small landlords may borrow for an individual property or project, while larger organisations, who expect to have an on-going development programme, are more likely to develop a long-term borrowing strategy aiming to take advantage of changing interest rates and the state of the housing market. A large landlord or developer might have a time-limited lending facility with a financial institution, allowing it to draw down funds as it thinks best over a set period of time and up to an agreed limit.

The same situation applies to private housing developers building housing for sale. They will have a lending and building strategy that allows them to take advantage of changing financial and market conditions. In addition, many developers will stockpile land, which allows them greater flexibility in the face of changes in demand.

It is not just the private sector that borrows to fund housing. The same applies to government at both the local and national level. Local government and municipalities may borrow to fund development, although their ability to borrow may be restricted by limits determined by central government. But central government itself borrows to fund public spending (including capital expenditure on housing) and welfare spending. While this borrowing might not be specifically for housing projects, it is certainly the case that many governments can only fund their current expenditure programmes if they are able to borrow from the financial markets.

So, we can say that housing systems depend on readily available long-term finance. Accordingly housing, which is by definition local, is dependent on a global financial network allowing borrowers access to the best deals. What further complicates matters is that the financial institutions that lenders and governments borrow from also trade in financial products themselves, and this includes the mortgages and loans of other financial institutions. It was this trading of financial products that led to the financial crisis of 2008. The quality of these financial products was dependent on both the ability of households to continue to make their regular mortgage payments and on the value of the houses used as security for the loans being maintained. If households default on their payments, due to unemployment or increased interest rates, and the value of properties falls to below that of the loan value, then the financial markets may cease to work properly and the sources of housing finance dry up. This happened in 2008, partly due to the fact that many of the financial products traded by international banks included sub-prime mortgages lent to US households with a poor credit history and an inability to fund their debts when house prices fell and interest rates rose. Accordingly, banks and other institutions found that the financial products they had purchased as a secure risk were actually worthless. So, the idea that trading these products was a way of sharing risk turned out in fact to be a means of spreading an infection.

What compounded this problem is that, while households borrow money over the long term – 25–30 years – mortgage lenders tend to borrow their finance over a much shorter period and typically for months rather than years. This means that lenders, and governments, will have to recycle their loans. Hence, there is the need for liquidity and confidence in the international financial system.

A final point is that, in order to work, these systems of lending and borrowing have been underpinned by government acting as the insurer of last resort against the risk of default and crisis. With owner occupation becoming dominant in so many developed countries, governments could not allow housing markets to fail. But the scale of the global market is now so large that no one government can hope to control it. At the time of the 2008 crash, some banks had bigger balance sheets than developed countries such as Spain, Ireland and the UK, and this made it very difficult for these countries to sort out their banking industries. So, while borrowing may allow us to gain access to the sort of housing we want, and developers to respond to our demands, this comes with risk and the threat of international economic insecurity attached. Never have the local and the global been so close together.

Think piece: The sub-prime mortgage market

In the first decade of the twenty-first century, many governments across the world encouraged owner occupation as a tenure for all households. This was possible because it took place during a prolonged period of economic growth which governments were regarding as permanent and demonstrative of a new economic era.

It is clear with hindsight that this was a mistaken view and dependant on continued growth and popular confidence. What broke this perception was the collapse of the US sub-prime mortgage market in 2006. These mortgages were sold to households with poor credit history, and often with little in the way of existing assets. Niall Ferguson (2008) refers to them as NINJA loans: no income, no job, no assets. Lenders in the US since the 1970s had been encouraged by government to lend to low-income households, and Fannie Mae and Freddie Mac were likewise encouraged to purchase these mortgages and so give them explicit support. Indeed, the role of these institutions changed from purchasing only high-grade securities to the active encouragement of sub-prime lending.

What made these mortgages attractive to households was their initial low rate of interest. This made the loan affordable in the short term, but there was a sting in the tail, in that the interest rate was scheduled to increase after the initial period. Perhaps households were persuaded that they could remortgage their house before the period ended and so keep on with the low rate, and this might have been possible if house prices continued to rise in the US. However, 2006 saw falls in prices in some cities which made remortgaging impossible and so households found themselves with higher repayments on what was now a declining asset.

This situation was compounded by the nature of the US mortgage market and particularly the fact that lenders tended to fund their lending through the sale of assets (i.e. mortgage debt) rather than through the deposits of savers, as was the case in other countries. The US mortgage market grew in the 1930s as a result of the backing given by government agencies which offered securitisation to lenders. But it was this means of supporting lenders that actually allowed the sub-prime market to thrive. Lenders could pass on their risky NINJA loans to other investors in return for continued funding. These investors would be national banks such as Lehman Brothers based in New York, who were happy to take the higher interest that came with these riskier loans. Mortgage brokers were, of course, more than happy to pass these loans on to others and continue with their lucrative trade.

In turn, the banks would then sell this risk on in the form of collateralised debt obligations (CDOs). This involved the combining of sub-prime mortgages with other less risky loans and selling them on to investors. These CDOs were often given the highest rating from ratings agencies like Fitch, Standard and Poor's and Moody's on the basis that they contained at least some high quality debt. The belief was that the use of these financial products would share the risk across financial institutions and so was of benefit to global financial markets. Sharing this debt around meant that the liability of any one institution would be limited.

The problem, however, was that these financial instruments were based on rather dodgy foundations. This was not readily apparent whilst house price continued to rise and sub-prime borrowers could afford their repayments. But, once house prices in cities like Detroit began to fall, the sub-prime market itself collapsed and it became evident that these debt obligations were potentially worthless and that the sharing of risk was not a means of protection but a form of infection. Instead of only a few local institutions being affected by mortgage default, the ripples were felt across the world.

What this shows is that lending is based on confidence: a bank borrows money because it thinks it will be paid back; a house is worth $200,000 because the buyer and seller believe it is. But, if confidence collapses, then a once-safe loan can appear to be very risky, and housing can all too readily become a liability.

Reference

Ferguson, N. (2008): *The Ascent of Money: A Financial History of the World*, London, Allen Lane.

Further reading

Shiller, R. (2008): *The Sub-Prime Solution: How Today's Global Financial Crisis Happened and What to Do about It*, Princeton, NJ, Princeton University Press.
Sowell, T. (2009): *The Housing Boom and Bust*, New York, Basic Books.

Discussion points

1   Can you imagine a housing system without borrowing?
2   How does the need to borrow influence housing markets?
3   Should government support the private borrowing of households?

See also

Further reading

King, P. (2009): *Understanding Housing Finance: Meeting Needs and Making Choices*, 2nd edn, London, Routledge.

King, P. (2010b): *Housing Boom and Bust: Owner Occupation, Government Regulation and the Credit Crunch*, London, Routledge.

Oxley, M. (2004): *Economics, Planning and Housing*, Basingstoke, Palgrave.

Part 6
Control

29 Control

Housing is by definition local. Its effects are experienced by those who live in or near to it. It is also something that is intensely personal: it is our home and we choose to make it into something that is unique to us. It is a place of privacy and intimacy, where we separate ourselves from the wider world. Yet, housing is an issue that government has always sought to manage and control. Government has never been happy simply to allow households and local communities just to get on with providing their own housing. There is a very long history of planning restrictions which limit what might be built and where it might be located, and there is an equally long history of attempts by government to tax housing and land. More recently, governments have supported the building of housing through subsidies to landlords and helped households to fund their housing costs through housing allowances or tax relief to offset their mortgage interest payments.

Clearly, one aim of subsidies is to encourage an activity. But, just like planning and taxation, they can also be used as a means of control. Once landlords and households become dependent on subsidies, government is able to exert some control over their behaviour and to direct them to meet broader social and political objectives. Government will target subsidies to particular priorities or to certain groups and it can increase or withdraw subsidies to encourage or discourage certain actions. For example, the withdrawal of subsidies to landlords might lead to higher rents, or the use of improvement grants might encourage existing housing to be renovated rather than demolished. Of course, this only works so long as government provides subsidies. There may be a temptation for government to reduce subsidies to landlords and take advantage of the fact that housing generates an income in the form of rent. So landlords might be forced to charge higher rents and thus allow government to reduce its subsidies. But it can only have any effective control over rent levels if there is still some subsidy to withdraw. So it is important to realise that, if governments did not provide subsidies to landlords or households, it would find it much harder to exert any control over housing systems, and this might affect its overall priorities.

Of course, it is possible to devise bureaucratic or statutory means of controlling rents but these often have perverse consequences, largely because they reduce the ability of landlords to make a reasonable rate of return on their investment. As something of an aside, it is interesting to note that governments do tend to prefer financial controls to legal ones. Governments are capable of using direct regulations and statutory controls. But this can often lead to complex and lengthy legal challenge. In addition, in many countries there is long history of local democracy and autonomy. So, in these cases, central government might be reluctant to impose its will too overtly, choosing instead to use the indirect, but equally effective, method of financial control.

But the fact that local housing is often owned or administered by local government or agencies means that there might be political conflict between the centre and locality. It is not uncommon for one political party to be in control at the centre, while others dominate over parts of local government. While it might be healthy for a democracy to have a plurality of political control, it does not help central government to implement its policies. What may well happen therefore is that central government accretes more power in order to control local government and thus ensure that they fulfil policies in the manner that central government requires.

Housing, of course, is not the only area that government supports financially, and these other areas might be more important politically. We might suggest that in many countries social housing is not a key election issue, unlike other areas of government like health care and old-age pensions. Social housing is means tested and provided for a selected group, and this means it tends to garner less support than services provided for all such as health care. Indeed, it might be the case that social housing is actually seen as a political liability for government, particularly if the tenure is linked to perceptions of crime, unemployment and anti-social behaviour. Accordingly, the primary concern for government will be to control social housing rather than encouraging it or letting it develop as it will.

A less controversial reason for control is that building houses is expensive. Landlords may have to borrow to fund their development and so will be paying off their loans for many years. Government may wish to control this borrowing to ensure the viability of landlords and to ensure that they do not become a liability to the taxpayer. But, once the houses are built, they will last for many years and will need funding for repairs and improvements. This can affect the need for subsidy as well as the rents that landlords wish to charge.

A final reason why governments seek to control housing is precisely because their previous attempts have failed. Much of government intervention is directly as a result of the failures or unintended consequences of past intervention. Partly this is due to the political cycle, in which one party takes over from another and wishes to undo or radically alter its predecessor's policies. But it is also because of the sheer complexity of housing systems, which means that government intervention is often poorly targeted and based on incorrect assumptions. We might think that this would lead politicians and policymakers to conclude that control is bound to fail and so withdraw. Yet, for precisely the reasons that we have discussed above, the response to failure has not been to increase local autonomy, but instead to increase the level of direct control.

Discussion points

1 What is the purpose of government subsidies – to encourage activity or control it?
2 Is housing a local or a national issue?
3 Are housing systems simply too complex for governments to control?

See also

Chapter 14 Social housing
Chapter 22 Sources of finance
Chapter 30 Government
Chapter 34 Planning

Further reading

King, P. (2006b): *A Conservative Consensus: Housing Policy Before 1997 and After*, Exeter, Imprint Academic.
Malpass, P. (2005): *Housing and the Welfare State*, Basingstoke, Palgrave.

30 Government

Government is the executive part of the state, which seeks to lead and direct the activities of that state. This implies that it is the most powerful element within a state, and this may indeed be the case. However, this does not mean that its role is not challenged, or indeed that government always gets its way. There are other sources of electoral legitimacy within the state such as the legislature and regional and local government. The actual role taken by government will differ according to the particular state. This means that we cannot be too specific here in detailing what government is concerned with. However, we can point to four general functions undertaken by government.

First, central government can plan and set the policy agenda. This might be through producing policy documents or commissioning and undertaking research, but also through the broad sweep of government fiscal and monetary policy. Second, the government can use its relationship with, or control over, the legislature. Most governments will have a legislative programme, which they seek to push through the legislature. In some countries, such as the UK, this is reasonably straightforward, assuming that the government has a sufficient majority in Parliament, but in some other countries that have separation of powers, such as the USA, this might be more difficult, especially when a president is coming towards the end of his second term and so cannot seek re-election. Third, government can attempt to regulate other bodies on the basis of its policies to ensure that they are implemented and its commitments fulfilled. Finally, government provides and directs finance towards particular policy objectives.

We can generalise, then, that there are four stages of government action, starting with planning, moving to legislating, then regulating and financing. However, it would be rather too simplistic to suggest that these are all straightforward processes and that one follows on from the other. In particular, there is some overlap and interaction between the various functions. The provision of finance is often dependent on regulation, and new legislation might be proposed because of the failure of regulation.

Using these four functions, we can point to a number of general actions that government can undertake that would be difficult for any other body, be it a market or an individual, to achieve. These actions might not apply to all states and, even where they do, we might not consider them all to be essential or necessary. Instead, these are actions that are quite commonly undertaken by government.

First, government can attempt to control and direct the economy by persuading the legislature to raise taxes and through its ability to set interest rates and through its own spending. Government spending in developed countries tends to be between 30 and 50 per cent of gross domestic product, and so changes in this spending can have a considerable impact on the economy as a whole.

Second, we can suggest that central provision might lead to economies of scale, in that large, nationally organised bodies can have considerable market power. In most countries the armed forces are the main, if not sole, purchaser of defence equipment, and so they can seek to drive down costs on the basis that suppliers might find it hard to sell their goods elsewhere. Likewise, the National Health Service in the UK, which provides 90 per cent of health care, is by far the largest purchaser of drugs in the country and so it has a considerable influence on drug prices in the UK. A decision not to allow the use of a drug in the NHS means that there is virtually no market for it. When this is allied to the government's role in regulating drugs, we can see that they can have a considerable effect on the prices that drug companies can charge, with the hope that this makes provision cheaper than would otherwise be the case. The downside, of course, is that the NHS is effectively the only market for drugs in the UK and so there is little in the way of competition to drive down prices: the issue therefore is whether government is capable of setting price levels properly.

This takes us to the third particular power of government. Just as government can raise taxes and make laws, it can override markets to allow for certain political targets to be met. It might be that society believes that issues, such as social justice or equity, outweigh a market-efficient outcome. Ensuring that all members of society are well housed might be seen as more important than consumer choice or free competition. Related to this is the ability that government has to take a national overview of spending and market activity and so attempt to balance competing objectives. Hence it is able to balance society's priorities for housing against those for health and transport, as well as for low taxes. Of course, a government may not achieve this, and indeed its particular priorities might be contested: young drivers might place transport as a higher priority than the elderly, who might want more money spent on health care.

Fifth, governments might seek to attain universality in provision. It might be that certain goods and services are seen as so fundamental that they should be provided to all relevant persons at the point of need. Again, the NHS is an example of this, but we can also point to old-age pensions, schools, income maintenance and housing allowances as examples. Likewise, government can attempt to achieve some form of uniformity of provision across the country. We might feel that certain forms of provision should be available to all equally; for example, a postal service may charge a standard rate for the delivery of a domestic letter regardless of the distance it has to travel. Likewise, paying a claimant's rent through a housing allowance system allows people to live in high-rent areas. Lastly, government can try to ensure that provision is at the right level and thus deal with market underprovision. In particular, a market might not provide housing for some minority groups or for people with disabilities whose needs increase the cost of housing considerably (providing wheelchair access, grab rails, accessible showers, etc.).

So there are a number of things that government can claim to do that would be difficult for any other body to achieve. In particular, it can be argued that a market will fail to do some or all of these activities. The point of dispute, of course, is whether all of these are necessary and are worth the other elements that come with a powerful central government. For example, uniformity in the postal service is one thing, but uniformity in terms of housing policies across an entire country might be another. Local populations differ, as do house prices, the scope of travel to work areas and local political priorities. For central government to impose a standardised structure of housing across the country might therefore be both unpopular and inefficient.

Think piece: The role of government

In some countries the state, normally at the local or municipal level, actually builds housing. In other countries, however, governments do not do this, but provide subsidies to private companies or charities to build housing.

How can we account for this difference? Is it cultural, historical accident or are there technical reasons why in some places governments build but not in others?

Discussion points

1   Is the government the most important player in housing systems?
2   What can government do that no other agency can?
3   If government is so powerful why does it often fail to meet its objectives?

See also

Further reading

King, P. (2006a): *Choice and the End of Social Housing*, London, Institute of Economic Affairs.
Malpass, P. (2005): *Housing and the Welfare State*, Basingstoke, Palgrasve.

31 Accountability

Some housing is provided using national or local subsidies for the purpose of housing local people. But it is unlikely, and indeed certainly unwise, that this money be spent without there being any strings attached. Those who helped to fund the housing, and those for whom it was ostensibly built, ought to have some say in the manner in which is built, allocated and subsequently managed. In particular, the landlords who build and manage the stock should be accountable to others who are deemed to have a stake in their provision. In certain situations the process of periodical democratic elections might provide for this accountability: if we do not like how the housing stock is being managed we can vote out one set of representatives and replace them with another.

Yet this might not work in practice. There might not be a majority view in a particular locality, only lots of competing minority views. In any case, getting rid of a majority on the local council might not be easy and may take several years, if it is possible at all. In addition, whilst social housing is expensive to build and the way it is allocated might be controversial, most local people will not be eligible for a particularist service based on need and vulnerability. Housing is often not the most important issue in an election, with other issues bearing more heavily on voters' minds. This will be the case particularly where only a minority of local people are, or can expect to be, tenants. So, if we cannot be housed in it, why should we bother how it is run? But this might mean that elected officials and their appointed managers are able to exercise a large amount of control over how services are run. Perhaps we should expect this though, as local managers not only have control over resources, but also will have a level of knowledge and skills that will outweigh those of the vast majority of local electors.

But central government has often provided much of the resources to build and manage social housing. It therefore requires some say in how local housing services are provided and run, so it can justify its expenditure to taxpayers and ensure that any money is well spent. Thus, social landlords have to be accountable to central government, as well as the local electorate.

The control of services by professionals is a very real issue, and is often referred to as producer capture. Professionals have the expertise and detailed knowledge that allows them to control the provision of services and so allow their interests to dominate. This domination can come about for a number of reasons. First, many providers are monopolies or exert a dominance over the provision of services. This is the case with social housing, in that there will most likely be either one landlord much larger than any others, or a situation where most landlords in an area co-operate rather than compete with each other. Monopolies do not have to have such regard for the needs of consumers, who are unable to go elsewhere for the service. Second, public services tend not to have valid indicators of performance. They do not make a profit or a loss and they are not competing for market share. So, again, they are not being tested in terms of their level of service. Third, because public bodies are often so large, they face problems in co-ordination and diseconomies of scale. This might be less of an issue with local social landlords, but some municipal authorities are very large, with a number of dwellings nearing the 100,000 mark. This size of organisation is difficult to control and to give the impression that each tenant or applicant matters. In addition, some large providers are now national bodies, and this might cause co-ordination problems between the central administration and local offices and agencies. As a result of these problems the issue of the accountability of social landlords has become increasingly important, with landlords having to justify their role and show that they are efficient and effective in their service delivery. So how do we measure whether social housing is doing what it should be doing?

The first issue is to decide just who is best able to answer this question. Is it society itself? Is it the government, which is deemed to be acting on society's behalf? Or should it be the legislature, which is meant to hold government to account? But should we not also take into account the voice of the users, who are the ones with the day-to-day experience?

The obvious way in which social landlords are held accountable is through the preparation and public presentation of financial accounts. This is where they record income and expenditure, and assets and liabilities for a given period. These will be audited independently and presented for public scrutiny. In order to achieve this, an organisation will undertake continuous recording of financial transactions. Accounts can be used to prove that income has been used properly, and that money has not been wasted or used frivolously.

Yet we need to remember that all this information is historical: the money will already be spent by the time we see the accounts. Therefore, in addition to accounts, an organisation will need to have financial regulations. These are a set of rules and policies which determine ways in which money can be spent and which will often be backed up by a system of internal audit to ensure that the rules are followed.

But there are other ways in which housing organisations can be made accountable. One way of ensuring accountability is through holding important meetings in public so that local people can see how decisions are taken and how their representatives behave. In addition, boards of management can be made to represent their local communities in terms of ethnic diversity and can include tenant representatives. Likewise, landlords can seek to ensure that their staff is representative of their communities in terms of gender and ethnicity. Finally, housing organisations may have to show that they are following government policies and are open to government scrutiny through a range of public bodies specifically appointed for that purpose. Government may appoint a body to regulate those housing organisations that seek public funding.

But the issue goes beyond formal lines of accountability and must also be concerned with how we measure whether an organisation is doing what it claims it does. Accountability should not just be a paper exercise, but should involve direct challenges to processes. In a market we might suggest that things such as demand, price and profit can be seen as measures of effectiveness. In this way we can say that consumers are getting what they want at a reasonable price, and that private businesses are doing well for their shareholders and staff.

But social housing is not simply a business, and there are other concerns that we need to include, particularly regarding social and welfare issues. Therefore profit is not a particularly relevant consideration, and indeed choice might not be as important in social housing as in markets. We can explore this further by looking at the concept of value for money.

This is essentially a means of determining that money is spent well and that it meets key aims and outcomes. Money will always be limited and thus decisions will have to taken between competing interests. For example, do we prioritise provision for the elderly or for young single people? Do we improve one estate before the other, and on what basis do we decide? In addition, there are always likely to be a range of stakeholders – applicants, tenants, staff, the local community, the taxpayer, etc. – and so decisions have to be made about whose priorities should prevail.

This raises the concept of opportunity cost. This is where the cost is conceived as the next best alternative opportunity forgone: if we had not decided to use the money for the agreed priority, what could we have done with it? As an example, we might judge expenditure on new social housing in terms of how many hospitals or schools we could have built with the money. So we can judge the manner in which we use resources by considering how else we might have used them.

What this suggests is that achieving value for money involves making the 'best' use of resources. This can never be an exact science, but there are four means of measuring value for money. First, they point to the concept of efficiency. This is determined by considering the relationship between inputs and outputs. Efficiency relates to quantity – if we increase inputs, do outputs increase by at least as much? – and also to quality, in that we are interested in the quality of the houses we build and not just how many.

Efficiency is often used as a means to measure how well markets work, but we can question how far it fits in terms of welfare and socially based organisations. Judgements about efficiency can prove to be controversial, largely because of difficulties of comparison. For example, how do we compare the efficiency of a small association concentrating on supported housing and a large national association that provides a full range of services? Their management costs will differ as will their staff/dwelling ratios. But, just because the smaller association appears to be much more expensive per dwelling, does this mean that it is doing a worse job than the larger, apparently more efficient association?

This brings us to the concept of effectiveness, which is where we are concerned not merely with the quantity and quality of outputs, but also with their impact. It is where we seek to assess how outputs contribute to the key objectives and expectations of those with a legitimate interest. A problem here is that these interests may clash, and so we need some means of prioritising them. This relates back to the general issue of who identifies the opportunity cost and how we arrive at some consensus on this.

Another problem, particularly when dealing with long-lived assets like housing, is that we cannot measure effectiveness immediately, but only over time. As an example, tower blocks built in the 1960s might have dealt with an immediate shortage of housing, but we might now question their effectiveness because of their higher maintenance costs, high voids and their general unpopularity. It therefore might now represent value for money to demolish them.

Third, there is the issue of equity. This relates to who benefits from a service and where the burdens of paying for it fall. Public bodies have to be demonstrably fair in how they use resources. However, not all households are at the same level in terms of income and opportunities. Therefore, allocating resources equally will not necessarily be fair, as landlords are trying to do different things for different needs groups. Therefore instead of a simple equal division of resources, we might need to take what might be called a 'value-added' approach that takes a detailed look at the impact of a landlord. Some landlords mainly house able-bodied households in general needs housing, while others might specialise in warden-supported accommodation for the elderly or those with physical disabilities. To return to the example of the small specialist association discussed above: how do we measure the added value of a small number of expensive dwellings to a relatively small, but highly dependent, client group?

Also related to the issue of equity are the means we choose to fund activities such as major repairs and improvements to the existing stock: do we charge the residents benefitting, or share the cost across the whole stock? Is it equitable that all residents, many of whom do not receive any additional benefit, pay for the improvements on only one estate? But, then, would it be fair to apportion the full cost to residents on that estate, many of whom might be unable to afford the increase?

The final means for determining value for money is by looking to experience. This relates to the expectations of the users and whether or not they perceive an improvement in the service. We might see this as the ultimate test, and in markets this can be easily measured by effective demand. In the public sector this might be measured by surveys and market research, but not so readily by changes in behaviour. One means of judging this in social housing might be through issues like voids and turnover, which can be taken to be indicators of poor quality.

What the issue of value for money does not settle, however, is the question of whose perception is dominant. Should we take the voice of the tenants as being of greater importance than the local community or the taxpayer, as represented by government? Being the dominant funder, the government often is the significant voice. Indeed, even when there are attempts to 'empower' tenants and give them a voice, this is a requirement imposed on social landlords by government. The flow of accountability is therefore very much upwards to government, rather than downwards to tenants. It is indeed the case that the body that controls resources and that has the ability to set the rules will demand that landlords be accountable primarily to them.

Think piece: Whose say matters most?

We can make a case for social housing belonging to different people, or rather acknowledge that there are many interests in the housing stock. But whose view matters most?

- Government might have provided some or all of the money and it is usually charged with meeting the main political, economic and social objectives of a nation, which would include housing provision for all citizens.
- Some of the funding might come from private lenders, who seek a reasonable rate of return on their investment.
- The housing is embedded within a local community and so should they not determine what is built and who gets priority when it is allocated?
- The landlord formally owns the properties and has to manage and maintain them. As the owners, shouldn't they have the largest say?
- For the residents, the dwelling is their home, where they live and raise their children. How can anyone's view be more important than theirs?
- How would you seek to reconcile these competing interests?

Discussion points

1 Does accountability really matter? Should we not be more concerned with helping the vulnerable?
2 Who is the most important stakeholder in a housing system?
3 Does equity matter more than efficiency?

See also

Chapter 6 Social justice
Chapter 18 Fairness
Chapter 22 Sources of finance
Chapter 30 Government

Further reading

Garnett, D. and Perry, J. (2005): *Housing Finance*, 3rd edn, Coventry, CIH.
King, P. and Oxley, M. (2000): *Housing: Who Decides?*, Basingstoke, Palgrave Macmillan.

32 Reform

No housing system is perfect, and there are always calls for reform. We can always point to policies that do not work as they should. In a world of finite resources and competing demands on these resources there will always be demands to do things differently. And, of course, things do go wrong and so there may in fact be a real need for change.

A number of the reasons for reform are clear-cut and obvious. For example, reformers may argue that the quality of housing needs to be improved, in terms of living conditions, standards of amenity and space standards. Similarly, it might be argued that there is an insufficient quantity of housing and we need therefore to build more, or that housing is currently too expensive for some or all households and needs to be made more affordable. These calls for reform ought to be backed by evidence and they ought to be empirically justifiable. Of course, in practice we know that these issues are rather more complex; for example, many developed countries have a crude surplus of dwellings and therefore objectively need not build any more. But we know that the immobility of housing and the segmentation of housing markets often means that we cannot rely on crude numbers when discussing shortages in either the quantity or quality of affordable dwellings.

It might be the case that in some countries the majority are already well housed, and so the housing crisis is defined instead as relating to specific unmet needs. These needs might be due to demographic changes such as an ageing population or immigration, or it might be because a society has become more sensitive to certain issues with the effect that needs that were hitherto hidden are now, as it were, discovered. This might relate to issues such as sexual abuse and domestic violence, which societies are now less likely to tolerate or ignore. The result is that the needs of abused children and women now become much more apparent and need to be addressed in ways that more general policies are incapable of doing.

A further justification, but one often used more widely, is the need to ensure that households become and remain independent. This may relate to the needs of the elderly and those with some form of disability, but it increasingly is used as a critique of the current structures of housing and welfare which are said to encourage a dependency on benefits and public provision and offer no incentive for households to move into employment or private provision. What we have is the apparent claim that services, which are a result of past reforms, are actually creating or sustaining problems such as poverty, unemployment and dependency. This adds a considerable degree of complexity to the call for reform. It shows that many reforms are deemed necessary to undo the apparent problems created by past reform, and, if it has gone wrong in the past, it might also go wrong again. Accordingly what is demanded here is not more action, but less. What reformers may be asking for is that government stops doing something that it currently does, whether it be paying benefits in the current configuration, or even building and renting houses at all.

This raises an important point that is worth dealing with here. Many calls for reform involve the call to do something additional to what is currently offered. A higher level of provision is needed, or a greater level of sensitivity is required to deal with particular groups who have special needs. However, for other reformers, the problem is that any provision is made at all. They might argue that government provision, which lowers the cost of housing for certain groups or for certain types of accommodation, crowds out private provision by creating market distortions and perverse incentives. It is therefore by no means the case that what is needed is more action. The reforms might actually involve a withdrawing of agency in a particular area, which would then allow other agents and institutions, including the households themselves, to provide out of their own resources.

This raises the vexed issue of motivation for reform and whether there are underlying motives, such as ideology or a vested interest. It might be that what motivates calls for change is not necessarily the empirical evidence but a preconceived ideological imperative. One often hears complaints that a political party of the left or right is only proposing something because it fulfils their ideological aims. It therefore can be stated that these changes are meant merely to fulfil a particular end that might not be as simple as more or better quality housing.

Likewise, it might be argued that a policy is aimed at meeting the interests of landlords at the expense of tenants, or vice versa; or a policy might be seen as benefitting a particular financial or propertied interest. What matters here, and why this can be a damaging claim if it becomes generally accepted, is that such policies are seen to benefit one group at the expense of another. The policy is not intended to have a general impact, but is rather aimed at promoting a particular interest or section within society.

It would be exceptional for the proponents of a policy accused of being either ideological or sectional (or both) to admit the case. The accusation is thrown at them from their opponents (who, it ought to be said, might be wishing to promote a particular interest of their own, it being the case that one's views are always acceptable and legitimate, while those of others are not), and the usual response is one of denial, outrage and resistance. The proponents of the contentious reform will instead suggest that their arguments are reasonable and based on evidence and are aimed at the general good. What tends to happen in these cases is that the issue moves away from the particular policy issue and becomes one of more general principle, often based on abstractions such as class, equality or fairness. In this sense, reform is about what sort of society we wish to live in, and so proposals to reform housing are actually part of a wider attempt to change society as a whole.

If one reason for reform is to deal with the policy mistakes of the past, this suggests that there is the possibility that reform will fail. Indeed, there is no real reason to assume that success is ever likely, bearing in mind the complexity of the systems being reformed. One key reason for policy failure is that the level of information available may well be limited. We will tend to have a very full picture of the current system, we know what parts of it function well and which do not, and we can readily point to what is wrong and in need of change. However, a proposal is still hypothetical, and so we can only have a limited understanding of how it might work in practice. It may be possible to successfully contrast known failings with hypothetical benefits, and this may be helped precisely by the fact that what is known to us is seldom seen as exciting or innovative. The consequences of the current failings are all too apparent, while the hypothetical system is all promise with nothing concrete to compromise it. Of course, this can work the other way as well, if the opponents of change are able to point to the comfortable parts of the current system and stress that departing from the known presents a considerable risk precisely because it is untested and unknown.

So it is by no means unreasonable and unusual to state that a system does not need reforming. We might wish to argue that conserving or preserving is a better aim. This might be because of its longevity or because of our fears for the future. It is quite common to suggest that, if something has stood the test of time, then it must have some value. On top of this, some might have a sentimental attachment to a particular way of doing things.

What this suggests is that there is always a risk with reform. While we can know the past and the present, we cannot know the future, and so there is always the possibility for error and misunderstanding. Having said this, there may be occasions where change is unavoidable and even the most cherished of institutions and policies are no longer tenable. Housing exists as part of a dynamic and interconnected social system and it cannot be insulated from the other parts of that system.

But this, too, means that no reform will ever be permanent: in time, it too will become outdated and unfit for purpose. This raises a final and very important point: despite the claims made by proponents of reform, perfection is not possible. This means that there will always be some part of the system that could work better. A system will always be capable of improvement. But, by the same token, we need to be aware that no reform will be perfect either, and that failure is as likely as success. This does not mean that creating a better system is impossible, but it might allow some to argue that it is.

Discussion points

1 What is the most important reason to reform housing systems?
2 How do you measure the success of reform, and who should be able to decide?
3 If reforms often fail should we not try in the first place?

See also

Chapter 5 Ideology
Chapter 21 Crisis
Chapter 30 Government
Chapter 31 Accountability

Further reading

Boyne, G., Farrell, C., Law, J., Powell, M. and Walker, R. (2003): *Evaluating Public Management Reforms*, Buckingham, Open University Press.

Part 7
Buildings

33 Development

The starting point for any discussion about housing development is to make the very apparent, but still important, point that housing is literally stuck in the ground. This might be obvious, but it is absolutely essential for understanding where much of the complexity of housing in terms of development, planning and economics derives. Virtually every other economic good (with the major exception, of course, of land) is portable. So, if there is a shortage of apples in Aberdeen, we can put some in a lorry in Kent and drive them up to Scotland. It is indeed possible to import apples from South Africa or New Zealand to the UK in a short period of time. Hence there will not tend to be much difference in the price of apples in Kent and Aberdeen. The same might not be the case for housing. If there is an increase in demand for housing in Aberdeen, we cannot move unwanted housing from somewhere else. We have to build more where it is needed and this means having the land, the finance and a willingness within the local community to accept this new development, which might be seen as a blight on the natural environment. So we really do have to take seriously the fact that most housing is stuck in the ground.

The next issue that we need to contend with is the need for land to build on. Unless we have the land, we cannot build. On one level there is no shortage of land in many countries. Or there would be if it wasn't already earmarked for other purposes such as growing crops, or society as whole did not feel that an amount of green space was worth keeping. Moreover, some land is simply not suitable to build on because it is not solid, sits half way up a mountain or is isolated from the necessary infrastructure. So there are competing uses to which land can be put, and there needs to be some means of allocating what is a relatively fixed resource (the experiences of the Dutch notwithstanding, it is not that easy to make new land).

But a further factor is that most, if not all, land will be owned by somebody, and that owner might not be prepared to allow it to be used for housing or any development at all. This might be because they wish to preserve it as it is, or perhaps because they feel that land values will rise in the future and so it will be profitable to hold off selling until a later date. Many countries have legislation allowing governments to compulsorily purchase land and property in some circumstances, but this often cannot be done without due process and or adequate compensation.

There will always be those who do not wish housing to be built in their area. They believe it adversely impacts on the environment or lowers their quality of life or the value of their property. They might see the need for more housing, but they may wish it could be somewhere other than near them. This is where development becomes politicised, as the interests of different groups collide. This is why there is the need for a rational system of land use planning. But it also shows that there is always conflict in housing development, and this is commonly between those who are already well housed and those who wish to join them in that state. The interests of the already housed may not coalesce with those seeking housing, and we need to remember that the former will always form a larger group than the latter.

The basic driver for development is demand and the assessment of the future need for housing. Obviously demand can be readily identified in terms of price signals and other measures, but future needs are dependent on certain assumptions about continued demand and changes in population. However, this points to another obvious but crucial point: it only makes sense to build housing where people want to live. While it is true that some countries have developed new towns and used subsidies and incentives to encourage firms and households to relocate, the norm is for development to follow demand. The majority of housebuilding will be in areas of high demand and where employment is relatively plentiful. This may involve calculating travel to work area and propensities to commute. For example, many people may work in major cities but live in satellite towns accessible by rail or road links. Thus, the demand for housing might not be within a city but within a certain easily commutable distance. This means that we might deal with housing demand in a popular area by building nearby. This might make economic sense, at least for a while, as land and property values will be lower.

There is little point building houses if there are no jobs locally or nearby, and the same applies if there is not the infrastructure alongside new development. This includes basics such as roads, water supply and sewerage, but it is also necessary to have shops, leisure facilities and other amenities. Some people may wish to live in the middle of nowhere, but they have to be able to get there and to turn the lights on when they arrive. So development does not merely concern the building of dwellings; it also necessitates infrastructure. The issue here is of who pays for the provision of these services. Is it a cost borne by the developer, who would then have to pass it on to the purchasers of the dwellings, or should it be borne by the local community?

There is always a risk involved in development and this is whether the dwellings can be sold and for a reasonable amount. The problem for the developer, however, is that virtually all of their costs are accrued up front. They do not start to recoup their costs until the development is finished. It might be possible to sell some properties off-plan or before the development is finished, and this gives some certainty to the developer, but it does not give them much of the money. This problem might be compounded by the time lag involved from putting in a planning application to the first residents moving in. Depending on the size of the scheme, this might take a number of years, by which time market conditions might have changed markedly. Developers therefore have an interest in forecasting the future state of housing markets and the economy more generally, and they need to plan the timing of their development accordingly. The result of this might be that developers are sitting on a considerable amount of land which they will only start developing once they sense market conditions are suitable. This might not, though, be very helpful for those households without suitable housing.

We have already seen that housing needs infrastructure, and this might impose a cost on society. But we also need to acknowledge that there may be considerable economic and social benefits in development over and above the new dwellings provided. Construction tends to be a labour-intensive industry and so it can have an effect on employment, and, because of the immobility of housing, this also means that the need for labour is local. Unlike an increased demand for goods such as cars or fridges, construction does not lead to the exportation of employment, but keeps it within the country in question.

Further to this is the significant derived demand that comes from house-building. There is, of course, the need for building materials and labour, but it will also lead to an increase in the demand for furniture and appliances, conveyancing, estate agents, painting and decorating, and other DIY materials and so on. Again this demand will be local rather than being exported abroad or to a different part of the country. So, housing development can have a considerable economic benefit, and this has led governments to support it over and above the need for housing.

While most economic goods tend to have a short life and will depreciate quite quickly, housing is a long-lived asset. Many people will live in houses that are older than they are, and it is not unusual for housing to last several hundreds of years. Not only does this mean that it will appreciate in value, but it also means that it will need repairing sometime during its lifetime. Even if the overall fabric of the dwelling is sound, parts of it will wear out or fail. But, more importantly, the expectations we have of our housing change over time, and so what was acceptable in terms of standards of amenity for our parents and grandparents may not suit us. Technological advances mean that new devices that were once luxuries become seen as necessities, and this means that we need more space to fit all these new machines in. The issue, therefore, is whether we attempt to pre-empt the inevitable increase in standards by building housing to a higher standard than we might currently expect, allowing it more readily to adapt to changing needs. However, the problem is that, while we know our needs will change, we cannot exactly predict how, and so we might not actually be able to plan effectively. While we know human beings age, we have not proven very good at predicting technological change and its impact. In addition, planning for the future will increase the cost of building now, and we might see that it is more important to keep building costs down and make housing more affordable. In this way we can build more dwellings now, even if we might have to adapt them at some future time (when, of course, it might no longer be our problem!).

The more technology we put into dwellings, the more complex they become. Indeed, we can see a house as a complex machine. It has lots of parts that we expect always to be working when we want them, such as heating systems, pipes, flushing toilets, showers, cookers, lighting and electricity. Much of the moving parts are hidden from view, behind walls, and so we lose sight of this complexity. At least we do until something stops working. We then come to realise just how complex the dwelling is, and we find that we need experts to repair and maintain it. We also find how expensive maintaining a dwelling can be, and as a result we may find it beneficial to insure the dwelling against damage and breakage, thus easing the cost of dealing with the unforeseen or an emergency.

Discussion points

1   Consider the broader economic benefits of housing development.
2   Does it matter that a house has become much more technologically complex?
3   Is it legitimate to campaign against a housing development next to your property?

See also

Chapter 12 Property rights
Chapter 23 Markets
Chapter 34 Planning

Further reading

Carmona, M., Carmona, S. and Gallent, N. (2003): *Delivering New Homes: Planning, Processes and Providers*, London, Routledge.
Golland, R. and Blake, R. (2003): *Housing Development: Theory, Process and Practice*, London, Routledge.

34 Planning

Most societies do not allow individuals to build just when and how they like, and this applies even if they own the land on which they are building. This is because building can impact on others in the vicinity and seriously compromise the quiet enjoyment of their own properties. Building a large extension to your house can block out some of the natural light to your neighbour's property, or it might mean that you can see directly into their house or garden and so invade their privacy. It might be that the building you propose to build is not in keeping with the general surroundings. If you live in the middle of a village consisting mainly of traditional houses made with locally sourced materials and you decide you want a five-storey mansion built mainly of concrete and glass, this might be felt to be inappropriate. It would be found offensive by other residents and seriously affect their daily lives. It might even adversely affect the value of their properties.

But we would also be concerned that anyone building a new property did so in a safe manner, both in terms of the actual construction and the finished product. We would want the building site to be safe for those working on it and living nearby, and we would want the building to be built according to some approved standards so that it does not collapse and injure or even kill passers-by and cause damage to the property of others. For all these reasons there may be planning restrictions placed on building, and it will be likely that permission will have to be sought from some statutory authority before building can commence.

In addition, a community might want to separate out certain activities, particularly to ensure that commercial and industrial activities are kept well away from residential areas. This is not merely to ensure that people are living well away from potentially hazardous materials and processes, but also because of more aesthetic reasons, to ensure that residential areas are pleasant and attractive places to live. This might also mean that planning authorities will restrict the numbers and type of dwellings that can be built.

There may also be more commercial reasons for zoning, in that there may be economies of scale in keeping certain activities together. Industrial areas might be kept to the edge of an urban area or close to a major road system. Residential areas of sufficient size make it viable for shops, bars and restaurants to thrive in the locality. But it also needs to be recognised that this zoning of areas can have an impact on the market value of the land. For example, designating a parcel of land for residential use may substantially increase its market value.

Such redesignation may, however, be resisted, particularly by those who feel they will be adversely affected by development. The same applies to new developments built in the proximity of existing dwellings. This has become known as Nimbyism (NIMBY: 'not in my back yard') and is a serious restriction on building, with local communities resisting new development even if the need for it has been clearly proven. The view may well be that people agree that new houses, or other development such as roads, landfill sites, factories, etc., are needed but they would rather they were built somewhere else. What this tells us is that planning involves coming to terms with conflicting and competitive interests, and some means has to be found for dealing with these divergent interests.

This leads us to the important point that, if land is used for one purpose, it cannot be used for another. If we build houses on some land, it cannot then be used for growing crops, or for walking through and admiring. Indeed, an awareness of the environmental impact of housebuilding has grown in recent years. This has led some countries to create so-called greenbelts or areas where no buildings of certain types are allowed. These may be around cities, providing a ring of green space and so limiting the expansion of urban areas. It might also be the case that building is forbidden or severely restricted in areas of particular natural beauty or where there is a special scientific interest being preserved. The restrictions placed on building in these areas puts more pressure on land in areas outside these zones and will tend to raise its value. Governments have attempted to deal with this by encouraging brownfield development, where building is on reclaimed land, or through redesignating buildings for residential use, allowing them to be converted from commercial or industrial uses. However, it needs to be remembered that any attempt to restrict development will have an economic impact and this will lead to pressures elsewhere.

So the largest part of the cost of building a house is the price of the land it sits on, and the difference in land values is one of the major contributory factors to the wide range in house prices that may exist within a country. However, house prices will also differ considerably even within a region or a city. This may be due to planning issues, but there are also two other causes. First, house types will differ, with some areas being more expensive and so more exclusive. Yet, even similar house types can vary wildly in value. This is due to the second issue, which is quite simply that the demand for housing exceeds supply. In some areas there is just not enough housing, whether it be for rent or to buy, and this has the effect of chasing up house prices. Planning restrictions, Nimbyism and the other issues that we have discussed might make this more of a problem in that they prevent or slow down any increase in supply. But we also have to recognise that the demand for housing is also tied into other issues such as employment opportunities, access to popular schools, perceived quality of life and other cultural opportunities. For example, certain types of jobs, such as in the arts and media, might only be available in large cultural centres and capital cities. Likewise, the demand for housing in an area might be increased because it is within the catchment area of a popular school, and the demand for housing in some villages might be high because it is a popular holiday location or because it has good transport links to a major urban area.

The fact that housing is stuck in the ground means that it is by definition local. That is why people get very annoyed when things are built right next to them: they cannot avoid them and cannot move their dwelling to somewhere else. So it matters therefore who is able to make decisions with regard to planning and building. In particular, should it be done nationally, where it is possible to gain an overview and set national priorities, as well as taking advantage of economies of scale? Or should planning decisions be taken locally based on local knowledge and the wishes of local people? The problem with local planning, however, is that decisions in one locality may affect what happens in neighbouring areas. For example, commuting patterns may not respect the jurisdictional boundaries of planning authorities, so that allowing the development of a major company's office in one location may increase the demand for housing in a neighbouring area. Taking decisions locally might therefore mean a lack of proper strategic thinking and not allow for all the relevant issues to be considered. But, on the other hand, it is also important that those with a direct interest in development have a say: after all they literally have to live with it.

The question, though, is just how local decisions are allowed to go: do we wish to restrict decisions only to those directly affected, and if we did would this not encourage Nimbyism and put the self-interest of a few over wider interests? We also have to acknowledge that some planning decisions will always be unpopular, such as the location of sites for dumping nuclear waste or for fracking. Likewise, the interests of certain groups such as Gypsies and Travellers, whose needs do not conform to the general pattern, might not be properly represented locally. Accordingly, we might want these most contested issues to be dealt with by a body better able to see both sides of the issue. There is need therefore to balance a range of interests both locally and nationally.

Discussion points

1   Why can't we simply build what we want?
2   How do we decide whether housebuilding is more important than growing crops or environmental issues?
3   Who should decide what is built in a local area?

See also

Chapter 12 Property rights
Chapter 23 Markets
Chapter 33 Development

Further reading

Carmona, M., Carmona, S. and Gallent, N. (2003): *Delivering New Homes: Planning, Processes and Providers*, London, Routledge.

Richardson, J. (2006): *The Gypsy Debate; Can Discourse Control?*, Exeter, Imprint Academic.

35 Architecture

Building houses is dirty, messy and very practical. It is literally 'hands-on'. Yet there is one person who might be involved in this business who sees themselves as an artist rather than a technician. This is the architect: the person who designs the building and provides the plans that others have to follow. Of course, architects must be competent in the technical aspects of building: they must know how materials interact with each other and how a house has to be fitted together. But an architect may not see this as enough for them. Designing little brick boxes is a rather dull activity for someone so highly trained and so they may feel that the need to innovate is irresistible. As an artist, they wish to express a particular aesthetic in their designs. They are concerned with how the building looks and what statement it makes to those people and buildings that are around it.

Good architecture, of course, is a delight, although we may not always be able to agree upon what this consists of. What some people consider to be an inspiring building may be considered an eyesore by others. What can appear challenging, fresh and original to one can be seen as an offence to another. In this sense, architecture is no different from any other art form. We disagree over a piece of music, a painting or a novel.

Yet there is a difference with architecture. If I do not like a picture I do not have to look at it again and the same applies to the piece of music that I can turn off and the novel that I can stop reading. But I cannot so readily avoid an ugly building if it is right outside my house or if I have to pass it on the way to work every morning. Architecture is a very public art form and we might suggest that this puts a particular responsibility on the architect to be civil. The word 'civil' has the same Latin root as 'city', and it relates to how we have to consider others whom we live close to and cannot avoid contact with. We need rules for a city, which make us 'civilised', and which show respect and 'civility' to others. So we might suggest that it is incumbent on an architect not merely to pursue their own aesthetic and or that of their client, but to recognise that everyone else has an interest in what is being built. An architect may be more concerned with how a building looks and what statement it makes, rather than whether it can function well a house. It is one thing to build a house according to some aesthetic principle or theory, but it is often another thing to have to live in it.

Most housing that is built is not architect designed, in the sense that each property has not been individually designed. It is rather more common for developers, be they public or private, to base their designs on standard plans, which are then tweaked to create a particular aesthetic that suits a locality or the whim of the developer. This has the advantage of reducing costs, although it might lead to a certain blandness and even standardisation.

Architects are more likely to be used by the wealthy for a one-off design, or by developers building prestige housing, again for the wealthy. There is the desire here for a more individual approach, and indeed people might pay a premium for a dwelling that is distinctive or built by a famous architect. But there is another call on the services of the architect. Much social housing has been built by public bodies and using public funds. This has encouraged some to use architects and give them the licence to innovate. This was particularly the case in the 1950s and 1960s with the architectural fashion for high-rise housing. This often involved the use of experimental building techniques and non-standard material based on a modernist aesthetic that high-rise living was the way forward.

But there is a key difference between these two forms of housing. Designing housing for the wealthy exists within a market where the clients have a choice. The client will either be the resident who has to live there or a developer who has to ensure that the dwellings can be sold. They will therefore ensure that the architect follows their instructions, or presumably they will go elsewhere. But in the case of public housing, the client – the public body – is under no market pressure and nor are the decision-makers likely to be living in the building. Instead the housing will be allocated to those in housing need who may well have no alternative. So we can suggest that those on low incomes were effectively being experimented on and had no means of questioning the outcome. The result has been some unpopular and even disastrous housing that proved to be uninhabitable, such as the Pruitt-Igoe estate in St. Louis, Missouri, which was lauded as a breakthrough piece of architecture when it was completed in 1956, but which soon began to decay and be deserted, and was finally demolished in 1972.

This does not mean that we should not have architects designing social housing. Rather, it suggests that the starting point for any development is the user who will live there and use the building. The aim should be to get the building to bend to the users, rather than the other way round.

Think piece: Le Corbusier and Algiers

In 1931 Algeria was still a French colony, as it would remain until 1962. The capital city, Algiers, was still in a largely traditional Arab form. As with many cities, there was overcrowding, poor sanitation and segregation. Accordingly, in 1931 it was decided to remodel the city. The famous and visionary Swiss architect, Le Corbusier, long a French resident, decided to put forward an alternative plan for the city.

In 1933 Le Corbusier presented his plan, which considered the removal of much of the traditional Arab architecture and its replacement with high-rise residential blocks, a new business quarter and a huge over-arching highway cutting through the city. Algiers would be converted from a traditional Arab city into a modern European-style metropolis. There would be little left of a local character, with the new Algiers appearing to be a city in the international style.

Perhaps unsurprisingly, Le Corbusier's plan did not fully meet with local approval. However, what is fascinating about this plan was that the architect did not present it as the basis for negotiation and consensus, but rather saw it as essential for the plan to be implemented as designed. Le Corbusier had a vision and insisted that it be implemented without compromise, and when he met with local resistance he lobbied the French government and even sought the support of the German-backed administration during the Second World War.

Le Corbusier's plan for Algiers was never built, but it does show something of the nature of architecture as art. Architects, even if most do not have the ambition, profile or sense of certainty of Le Corbusier, put forward a vision, and this is something often based on abstract principles. In this case, it was a vision of modernity that included the idea that traditional architecture was now redundant. But this can appear to be an imposition to those who love and live in a particular place. The issue, then, is how far we should listen to architects and what their proper role is in designing buildings for real human beings to live and work in.

Reference

Jencks, C. (2000): *Le Corbusier and the Continual Revolution in Architecture*, New York, Monacelli Press.

Discussion points

1 How important is it for a building to be civil and so fit in with its surroundings?
2 How do you involve the potential users in the design of social housing?

See also

Chapter 13 Desire
Chapter 14 Social housing
Chapter 33 Development
Chapter 34 Planning

Further reading

King, P. (2008): *In Dwelling: Implacability, Exclusion and Acceptance*, Aldershot, Ashgate.
Scruton, R. (1994): *The Classical Vernacular: Architectural Principles in the Age of Nihilism*, Manchester, Carcanet.

36 Space and place

Housing is part of something bigger. Much of our world has been created by ourselves. We have changed it, domesticated it where we can and built on it. We live in a world that has been largely created by ourselves, and housing is just a part of this. Even if we live in a tiny village in the deepest countryside, this remains a created environment, and it is still connected into a much larger human-made structure which we can call the built environment.

Many of us live in cities, albeit of varying sizes, and so we are used to our own relative insignificance. We are just one person traversing a large and impersonal urban landscape of roads, offices, shops, housing and so on. We seemingly cannot influence this world, but can only accept it and go with the flow. We have to follow the rules and understand the patterns of this urban world. But, once we have done so, we have much of what we want and need ready to hand.

Humans are relational beings. Our lives consist of relating to other physical entities, be they animate, as in other people, or inanimate like the environment. We operate within space as physical beings looking outwards into the world. Because of these relationships, we give space significance. We might argue that this creation of significance is what creates place out of space.

We can see place as bounded space, as something that is particular to us, that we can define in some way – which need not be by a physical boundary – and which we take as meaningful. This place is ours, it is familiar and we feel comfortable there. This place might be our housing, but it will also include the surroundings of our dwelling and the larger built environment in which it is located.

Our understanding of space and place, according to Ali Madanipour (2003), is conditioned by three things. First, there is the idea of spatial scale, going from our bodies through to the city. Second, there is the sense of openness, which relates to the degrees of exclusivity and openness that we experience, again ranging from the most private to the most public. Third, Madanipour refers to relationships and the different modes of social encounter from personal to impersonal.

We can link these three elements – scale, openness and relationships – together by imagining a series of circles around an individual, moving ever further away, becoming more transparent and porous as they do. We start with our own body and the private, interior space of the mind. This is entirely enclosed and open only to us. No one can find out what is in there without our mediation. If we keep ourselves to ourselves, it can be completely private, and inside our head there need only be a monologue with no need for a conversation.

Moving slightly further away, we have the immediate space around us, what we might call our personal space. There are only certain special people whom we are prepared to share this space with because this involves intimacy. We will seek to protect this space and keep others away. This need not be a space that we share, and we hope that we can keep the sharing mutual. This applies to the next circle, which is what we might call private space, or in other words the home. This is a place of privacy, where we can exclude most others. But it is also a place of sharing, caring and nurturing. This is the first level of space, then, that we share with others. These are people with whom we have much in common and so it can be a place of mutuality and support. But it can also be a place of conflict, particularly between generations. These conflicts might be over space, with the problems that sharing brings, but they also might be about relationships and the ability of members of the household to stake out their own identity and sense of place.

Outside the home is interpersonal space. This is not space that we can control. We do not own it. It is also the level at which we begin to come into contact with strangers, with people we do not necessarily know and whose lives we do not share. But this is still a place that is familiar and known to us. It is the road in front of our house, the walk to school, the local bus stop or our working environment. It remains part of our everyday experience and it may be that the people we see are familiar in the sense that we see them quite regularly. However, we do not share intimacies and have no real knowledge of their lives.

And then there is what we might call impersonal public space, or the city. This is anonymous and implacable to us. We cannot control it or mould it to suit ourselves. The city seems to exist and go about its business with little regard for us. It is full of people whom we have not seen before and will not see again. It is a space of polite but formal relations where we are guarded and do not show much of ourselves.

So we can see the built environment as a series of circles moving outwards from each of us. As the circles move away, they become porous and we cannot influence what is inside. I also have to recognise that the set of circles around me interacts with those of my family but also everyone else that I come into contact with. These circles are inhabited by other people with the same set of relationships with the environment as us. Our circles, so to speak, interlink and start to affect each other. This helps us to understand that many of the spaces we inhabit are used by others, and it may be that they feel as attached to them as we are. There are many places that feel as if they are mine, but that does not mean that I have any exclusive rights over them.

Think piece: Feeling at home

Sometimes we have to work hard to turn space into place. Paul Oliver discusses the distinctive architecture of the village of Akyazi in the Black Sea region of Turkey. He suggests that, whilst there is a considerable variety amongst the dwellings of the village, they are all different from the vernacular architecture of the region. Instead of the typical style of a timber frame with adobe infill, the houses in Akyazi are constructed on massive posts driven into the ground. They also feature a deep veranda structure, which is again untypical of other villages in the region. But what Oliver sees as being particularly significant about this village is 'The residents' pride in their Georgian ancestry, their defence of their culture and reluctance to relinquish the roots of their past, and the slow, century-old process of adjustment, which includes the adoption of the Turkish language' (2003, p. 55).

Akyazi is a village of Muslims of Georgian descent who originally settled in Turkey in the 1880s following the Russian defeat of the Ottomans and the reinstating of Christianity in Georgia. Akyazi is therefore the result of migration, but can now be seen as an act of remembrance, as a form of staying still in an unfamiliar environment. Oliver states that 'Its houses remain Georgian in design, construction and use, and though four or five generations have passed since its birth, its present builders cling tenaciously to their traditions, while erecting dwellings that are still varied within its norms' (2003, p. 55). Despite over 100 years of living in Turkey and the passage of generations, the villagers of Akyazi cling to their Georgian roots, even as they live and thrive within their adopted region.

These Georgians live in an environment that is not of their own choosing. Yet they have adapted it and sought to control what they can. This may be an extreme example, but is this not what all migrants seek to do when they settle in a new place: to make an old place out of new space?

Reference

Oliver, P. (2003): *Dwellings: The Vernacular House Worldwide*, London, Phaidon.

Discussion points

1   What turns space into place?
2   How important is it for you to control your personal space?
3   Why does the home sometimes become a place of conflict?

See also

Chapter 1 Housing and home
Chapter 13 Desire

Further reading

King, P. (2008): *In Dwelling: Implacability, Exclusion and Acceptance*, Aldershot, Ashgate.
Madanipour, A. (2003): *Public and Private Spaces of the City*, London, Routledge.
Scruton, R. (1994): *The Classical Vernacular: Architectural Principles in the Age of Nihilism*, Manchester, Carcanet.

Conclusions

A continuing conversation

Housing always looks like it is changing: it seems that so much is going on. But much of what we call 'housing' stays the same. What housing is, what it does, how we use it, the routines of provision, maintenance and policymaking remain largely the same.

But that which stays the same goes on and on. It does not end: once one household is housed, another one comes to the top of the queue; at one time we buy a house and at another time we sell it; we knock down so that we can build anew; what we think has been dealt with by sound policymaking comes back again; and so it goes on.

Housing is basic, but it is also difficult. We need a lot of it; we need it everywhere; and we need it all the time. Some of us are rather fussy about what we live in and where it is, and perhaps some of us are greedy and want rather more of it than we should. And we have to deal with the consequences of our actions, some of which might be rather selfish.

Housing is basic, but it is also expensive. Those who buy a house spend a third of their lives paying for it, and those who rent never stop paying. But what we are paying for is what we assume already to be ours. We use it knowing that it belongs to us.

My house is basic, but it is also mine. I feel that it belongs to me. Yet my wife and my daughters also refer to it as 'my house'. This does not diminish any sense of ownership on my part. In fact, it makes me very happy that my wife and daughters think that my house is also theirs. It is exactly what I had desired would happen. We possess housing both as individuals and as households, and it is this dual sense of ownership – which has nothing to do with tenure – that makes housing meaningful to us.

Housing is local. It is literally stuck in the ground. But housing is also global. We only need to think of where the money comes to pay for it and the mess that followed the 2008 financial crisis to realise this. But, whatever the mess, and whatever the grief that it causes us and to those around us, we have no choice but to return to our cosy little house, anchored firmly in the ground next to its neighbours, linked by roads, pathways and green space into the local community, the city, the nation.

What does not end – and will not end – is the talking about housing. The conversation about what we need, where we need it, who will pay for it, and who is first in the queue, will continue. It is a conversation that we can all join in and which we can all understand. And this is precisely because housing does not change that much at all.

Further reading

Having now read this book, you might want to continue the conversation with someone else. Here are a number of good starting points:

Atkinson, R. and Jacobs, K. (2016): *House, Home and Society*, Basingstoke, Palgrave.

This is perhaps the most convincing recent attempt to develop a sociology of housing. It links our experience of home with global economic influences to develop a lively picture of the world of housing.

Clapham, D. (2005): *The Meaning of Housing: A Pathways Approach*, Bristol, Policy Press.

This is now a staple of the housing literature, but it is an excellent means for exploring the connections between housing, employment, education and leisure. It follows our housing careers throughout our lives and shows why we always need housing.

King, P. (2008): *In Dwelling: Implacability, Exclusion and Acceptance*, Aldershot, Ashgate.

One of mine, but I hope that it is a good way in which to look at more philosophical approaches to housing. The book explores how we use our housing and what it means to us. The book uses examples from film and fiction to explore why housing is so important to us personally.

McNelis, S. (2014): *Making Progress in Housing: A Framework for Collaborative Research*, London, Routledge.

Sean McNelis's book is perhaps the most original as well as one of the most interesting books written on housing for a very long while. He uses what at first appears to be a daunting and complex social theory to produce what is a fascinating and coherent look at contemporary housing issues, mainly from an Australian perspective.

Smith, S. J., Elsinga, M., Fox O'Mahony, L., Eng, S. O. and Wachter, S. (2012) (Eds): *International Encyclopedia of Housing and Home*, Oxford, Elsevier.

This is a truly monumental work consisting of over several hundred chapters on all aspects of housing. It is truly international and comprehensive in its focus and so is perhaps the best place to start if you want to find out what is happening throughout the world.

Bibliography

Adams, I. (1993): *Political Ideology Today*, Manchester, Manchester University Press.

Albon, R. and Stafford, D. (1987): *Rent Control*, London, Croom Helm.

Bachelard, G. (1969): *The Poetics of Space*, Boston, MA, Beacon Books.

Blunt, A. and Dowling, R. (2006): *Home*, London, Routledge.

Boyne, G., Farrell, C., Law, J., Powell, M. and Walker, R. (2003): *Evaluating Public Management Reforms*, Buckingham, Open University Press.

Bradshaw, J. (1972): 'The Taxonomy of Social Need', in McLachlan, G. (Ed): *Problems and Progress in Medical Care*, 7th series, Buckingham, Open University Press.

Brown, T. and King, P. (2005): 'The Power to Choose: Effective Choice and Housing Policy', *European Journal of Housing Policy*, 5, 1, pp. 59–75.

Carmona, M., Carmona, S. and Gallent, N. (2003): *Delivering New Homes: Planning, Processes and Providers*, London, Routledge.

Clapham, D. (2005): *The Meaning of Housing: A Pathways Approach*, Bristol, Policy Press.

Daly, G. (1996): *Homeless: Policies, Strategies and Lives on the Street*, London, Routledge.

De Soto, H. (2000): *The Mystery of Capital: Why Capitalism Triumphs in the West and Fails Everywhere Else*, London, Black Swan.

Dorling, D. (2008): *The Population of the UK*, London, Sage.

Dorling, D. (2014): *All That is Solid: How the Great British Housing Disaster Defines Our Times, and What We Can Do About It*, London, Allen Lane.

Doyal, L. and Gough, I. (1991): *A Theory of Human Need*, Basingstoke, Macmillan.

Elster, J. (Ed.) (1986): *Rational Choice*, Oxford, Blackwell.

Elster, J. (1995): *The Cement of Society: A Study in Social Order*, Cambridge, Cambridge University Press.

Ferguson, N. (2008): *The Ascent of Money: A Financial History of the World*, London, Allen Lane.

Freeden, M. (1991): *Rights*, Buckingham, Open University Press.

Garber, M. (2000): *Sex and Real Estate: Why We Love Houses*, New York, Anchor Books.

Garnett, D. and Perry, J. (2005): *Housing Finance*, 3rd edn, Coventry, CIH.

Golland, R. and Blake, R. (2003): *Housing Development: Theory, Process and Practice*, London, Routledge.

Goodin, R. (1998): 'Social Welfare as a Collective Social Responsibility', in Schmidtz, D. and Goodin, R: *Social Welfare and Individual Responsibility*, Cambridge, Cambridge University Press, pp. 97–195.

Green, E. (2006): *Thatcher*, London, Hodder Arnold.

Griffin, J. (1986): *Well-Being: Its Meaning, Measurement and Moral Importance*, Oxford, Clarendon.

Harlow, M. (1995): *The People's Home?: Social Rented Housing in Europe and America*, Oxford, Blackwell.

Hayek, F. (1948): *Individualism and Economic Order*, Chicago, IL, Chicago University Press.

Hayek, F. (1967): *Studies in Philosophy, Politics and Economics*, London, Routledge.

Hayek, F. (1978): *New Studies in Philosophy, Politics, Economics and the History of Ideas*, London, Routledge.

Hayek, F. (1982): *Law, Legislation and Liberty*, London, Routledge and Kegan Paul.

Hayek, F. (1988): *The Fatal Conceit: The Errors of Socialism*, London, Routledge.

Hills, J. (1997): *The Future of Welfare: A Guide to the Debate*, revised edn, York, Joseph Rowntree Foundation.

Jencks, C. (2000): *Le Corbusier and the Continual Revolution in Architecture*, New York, Monacelli Press.

Jones, C. and Murie, M. (2006): *The Right to Buy: Analysis and Evaluation of a Housing Policy*, Oxford, Blackwell.

Kemp, P. (1997): *A Comparative Study of Housing Allowances*, London, HMSO.

Kemp, P. (Ed.) (2007): *Housing Allowances in Comparative Perspective*, Bristol, Policy Press.

King, P. (2003): *A Social Philosophy of Housing*, Aldershot, Ashgate.

King, P. (2004): *Private Dwelling: Contemplating the Use of Housing*, London, Routledge.

King, P. (2005): *The Common Place: The Ordinary Experience of Housing*, Aldershot, Ashgate.

King, P. (2006a): *Choice and the End of Social Housing*, London, Institute of Economic Affairs.

King, P. (2006b): *A Conservative Consensus: Housing Policy Before 1997 and After*, Exeter, Imprint Academic.

King, P. (2008): *In Dwelling: Implacability, Exclusion and Acceptance*, Aldershot, Ashgate.

King, P. (2009): *Understanding Housing Finance: Meeting Needs and Making Choices*, 2nd edn, London, Routledge.

King, P. (2010a): *Housing Policy Transformed: The Right to Buy and the Desire to Own*, Bristol, Policy Press.

King, P. (2010b): *Housing Boom and Bust: Owner Occupation, Government Regulation and the Credit Crunch*, London, Routledge.

King, P. (2012): 'The Ethics of Affordable Housing', in Chadwick, E. (Ed): *Encyclopedia of Applied Ethics*, 2nd edn, San Diego, CA, Academic Press, vol. 1, pp. 72–78

King, P. and Oxley, M. (2000): *Housing: Who Decides?*, Basingstoke, Palgrave Macmillan.

Levine, D. (1995): *Wealth and Freedom: An Introduction to Political Economy*, Cambridge, Cambridge University Press.

Madanipour, A. (2003): *Public and Private Spaces of the City*, London, Routledge.

Malpass, P. (2005): *Housing and the Welfare State*, Basingstoke, Palgrave.

McNaughton, C. (2008): *Transitions Through Homelessness: Lives on the Edge*, Basingstoke, Palgrave.

Miller, D. (1976): *Social Justice*, Oxford, Oxford University Press.

Mulder, C. (1996): 'Housing Choice', *Netherlands Journal of Housing and the Built Environment*, 11, 3, pp. 209–232.

Murray, C. (1985): *Losing Ground: American Social Policy, 1950–1980*, New York, Basic Books.

Murray, C. (1996): *Charles Murray and the Underclass: The Developing Debate*, London, Institute of Economic Affairs.

Niemietz, K. (2011): *A New Understanding of Poverty*, London, Institute of Economic Affairs.

Nozick, R. (1974): *Anarchy, State and Utopia*, Oxford, Blackwell.

Oliver, P. (2003): *Dwellings: The Vernacular House Worldwide*, London, Phaidon.

Oliver, P. (2006): *Built to Meet Needs: Cultural Issues in Vernacular Architecture*, London, Routledge.

O'Neill, J. (1998): *The Market: Ethics, Knowledge and Politics*, London, Routledge.

Oxley, M. (2004): *Economics, Planning and Housing*, Basingstoke, Palgrave.

Oxley, M. and Smith, J. (1996): *Housing Policy and Rented Housing in Europe*, London, Spon.

Power, A. (1993): *From Hovels to High Rise: State Housing in Europe Since 1850*, London, Routledge.

Rawls, J. (1971): *A Theory of Justice*, Oxford, Oxford University Press.

Rawls, J. (2003): *Justice as Fairness: A Restatement*, Cambridge, MA, Harvard University Press.

Richardson, J. (2006): *The Gypsy Debate; Can Discourse Control?*, Exeter, Imprint Academic.

Richardson, J. (Ed.) (2010): *From Recession to Renewal: The Impact of the Financial Crisis on Local Government and Public Services*, Bristol, Policy Press.

Robinson, R. (1979): *Housing Economics and Public Policy*, Basingstoke, Macmillan, pp. 55–56.

Saunders, P. (1990): *A Nation of Home Owners*, London, Unwin Hyman.

Schmidtz, D. (1998): 'Taking Responsibility', in Schmidtz, D. and Goodin, R: *Social Welfare and Individual Responsibility*, Cambridge, Cambridge University Press, pp. 1–96.

Scruton, R. (1994): *The Classical Vernacular: Architectural Principles in the Age of Nihilism*, Manchester, Carcanet.

Scruton, R. (2001): *The Meaning of Conservatism*, 3rd edn, Basingstoke, Palgrave.

Shiller, R. (2008): *The Sub-Prime Solution: How Today's Global Financial Crisis Happened and What to Do about It*, Princeton, NJ, Princeton University Press.

Smith, A. (1976): *An Inquiry into the Nature and Causes of the Wealth of Nations*, 2 vols, Indianapolis, IN, Liberty Fund.

Smith, S. J., Elsinga, M., Fox O'Mahony, L., Eng, S. O. and Wachter, S. (2012) (Eds): *International Encyclopedia of Housing and Home*, Oxford, Elsevier.

Sowell, T. (2007): *Basic Economics: A Common Sense Guide to the Economy*, 3rd edn, New York, Basic Books.

Sowell, T. (2009): *The Housing Boom and Bust*, New York, Basic Books.

Townsend, P. (1979): *Poverty in the United Kingdom*, London, Penguin.

Turner, J. F. C. (1976): *Housing by People: Towards Autonomy in Building Environments*, London, Marion Boyars.

Waldron, J. (1993): 'Homelessness and the Issue of Freedom', in: *Liberal Rights: Collected Papers, 1981–1991*, Cambridge, Cambridge University Press, pp. 309–338.

Wilkinson, R. and Pickett, K. (2009): *The Spirit Level: Why More Equal Societies Almost Always Do Better*, London, Allen Lane.

Index

Any index should be looked at in conjunction with the Contents page, but this is particularly the case with this book. The reader's first port of call to find a topic should be the Contents page. The index below only consists of entries additional to the main chapter titles. Where chapter titles are mentioned it is because there are important discussions outside of the main chapter.